電気電子系学生のための英語処方　改訂版

論文執筆から口頭発表のテクニックまで

English Manners for Electrical and Electronics Engineering Students, Revised Edition:

Techniques for Writing and Presenting Successful Scientific Papers in English

馬場 吉弘　著

William A. Chisholm　監修

電気学会

はじめに(初版)

　電気電子工学分野においては，質の高い論文を掲載する多くの英文雑誌が出版されています．また，毎年多くの国際会議が開催されています．このような状況のなか，電気電子工学系の学生や若手研究者・技術者が研究成果を英語論文にまとめ，英語で発表する機会も増えてきています．

　上手に英語論文を執筆し，英語で発表するためには，英語を母国語とする研究者・技術者やアメリカやイギリスに長期間住んでいる(あるいは住んでいた)研究者・技術者が執筆した論文を多数読み，また彼らが発表するのを何度も聞き，そこで使われている表現を真似ることが役立ちます．しかし，学生や若手研究者・技術者は，このような経験自体が少ないため，英語での論文執筆や発表の準備に多くの労力と時間を要します．たとえば，英語論文の執筆時や発表の準備段階において，以下のような疑問や不安が次々に生じ，順調に進められないのが普通です．

- 「題目は何文字ぐらいが適当だろうか？」
- 「能動態を用いてもよいのだろうか？」
- 「時制はどのように選ぶのだろうか？」
- 「冠詞はどのように選ぶのだろうか？」
- 「マクスウェル方程式は Maxwell's equations と書くのだろうか？ the Maxwell equations あるいは the Maxwell's equations と書くのだろうか？」
- 「ハイフン(-)やコロン(：)はどのように使うのだろうか？」
- 「発表はどのような言葉で開始すればよいのだろうか？」
- 「5 A は five ampere と読むのだろうか？ あるいは単位を複数形にして five amperes と読むのだろうか？」
- 「x^{-3} はどのように読むのだろうか？」
- 「数式はどのように読むのだろうか？」
- 「ポスター発表はどのようにすればよいだろうか？」
- 「論文査読状況の照会時にはどのような電子メールを送ればよいだろうか？」

　本書は，上述のような疑問に答え，さらに電気電子工学分野の読者に役立つように配慮して書いた英語論文執筆および発表に関するガイドブックです．本書の英文は，アメリカ生まれでカナダ在住の William A. Chisholm 博士(アメリカ電気電子学会フェロー)に校閲を受けていますので，読者の皆さんには安心して本書を読んでいただけると思います．なお，イギリス英語とアメリカ英語では表現の異なる場合がありますが，本書ではアメリカ英語での表現を採用しています．

　本書は，三つの章と付録から構成されています．**第1章**では，英語論文の執筆法について説明しています．また，能動態と受動態の使い方，時制の選び方，助動詞の使い方，冠詞の選び

は じ め に

方などについても説明しています．**第2章**では，英語論文の口頭発表法について説明しています．また，グラフや表の説明法，数や数式の読み方，質問への対処法などについても説明しています．**第3章**では，電気電子工学の各分野における英文教科書や学術論文に記載されている文章を引用し，役立つ諸表現の実例を和訳とともに示しています．**付録**では，論文投稿から掲載あるいは発表にいたるまでの流れ，電子メールの文例，アメリカ留学関連の手続きなどを紹介しています．

　最後になりましたが，本書を執筆するにあたり，有益なご助言を賜りました桂井誠委員長をはじめとする電気学会出版事業委員会の皆様，電力中央研究所の新藤孝敏博士，同志社大学の雨谷昭弘教授，長岡直人教授，本書の原稿を丁寧にお読みくださり，多数の貴重なコメントをくださいました閲読者の方に心から感謝いたします．また，学生時代に執筆した英語論文原稿を親身になって添削指導してくださいました東京大学の石井勝教授，アメリカ留学中に執筆した論文原稿を丁寧に添削指導してくださいましたフロリダ大学のVladimir A. Rakov教授に深く感謝いたします．最後に，本書の原稿を3年以上もの間辛抱強くお待ちくださり，また本書の出版に際してご協力くださいました電気学会編修出版課の皆様に，心からお礼を申し上げます．

2013年1月吉日

馬場吉弘

改訂にあたって

　本書の初版が出版されてから 10 年以上が経過しました．この間に，電気電子工学関連の研究にも人工知能（AI）の応用が進み，一分野を形成するまでに至っています．英文論文執筆時に便利な Grammarly などの英文校正ツールや論文および被引用論文の検索や業績管理の自動化に役立つ Google Scholar などの利用も広がりました．また，学術雑誌の影響力を示す指標のインパクトファクタや研究者のその分野での貢献度を単一の数値で表す h-index の認識も広がりつつあります．さらに，機械翻訳ソフトや生成 AI の利用による意図しない剽窃の発生を防止し，投稿予定論文と著者本人の過去の論文との重複部分を最小限にする目的で，論文投稿時には iThenticate 等のツールにより重複度チェックを行うことが一般化しています．不適切なオーサーシップの問題も顕在化しており，共著者の選定はこれまで以上に慎重に行うことが求められています．新型感染症の数年間にわたる流行により，以前はほとんど利用されることのなかったオンラインやオンデマンド形式での発表も行われるようにもなりました．

　以上のような状況の変化を考慮して，内容を追加更新し，大学院生，企業の若手研究者や技術者に役立つ最新情報も含む英語論文執筆・発表ガイドブックに改訂しました．具体的には，第 1 章「英語論文執筆法」に．第 1.9 節「論文投稿前の文法および剽窃チェックの仕方」を，第 2 章「英語論文口頭発表法」に第 2.7 節「オンライン発表の心得とオンデマンド発表資料の作成」を，第 3 章「電気電子工学分野における諸表現」に第 3.6 節「機械学習・人工知能（AI）に関する表現」を，付録に付録 3「h 指数について」を追加しています．また，不適切なオーサーシップについては第 1.1.3 節に追記し，生成 AI 利用による思いがけない著作権侵害の可能性や Grammarly などの英文校正ツールの利用については第 1.9 節で説明し，論文および被引用論文の検索や業績管理の自動化に役立つ Google Scholar などの利用については付録 3 で説明しています．なお，本書を 15 回からなる講義で教科書または参考書として利用される場合には，下記のように分けて説明されると，英語での論文執筆から口頭発表のテクニックまでを無理なく教授できると思います。この改訂版が読者の皆さんに役立つことを願っています。

- 第 1 回　「英語論文の構成　その 1」として，1.1 の 1.1.1 から 1.1.7
- 第 2 回　「英語論文の構成　その 2」として，1.1 の 1.1.8 から 1.1.14
- 第 3 回　「能動態と受動態の使い方，時制の選び方」として，1.2 と 1.3
- 第 4 回　「助動詞の使い方，冠詞の使い方」として，1.4 と 1.5
- 第 5 回　「接続詞，分詞構文，関係代名詞，前置詞の使い方」として，1.6 と 1.7
- 第 6 回　「ハイフン，コロンなどの使い方，投稿前の文法と剽窃チェックの仕方」として，1.8 と 1.9
- 第 7 回　「発表用スライドの作成法と発表法　その 1」として，2.1 の 2.1.1 から 2.1.6
- 第 8 回　「発表用スライドの作成法と発表法　その 2」として，2.1 の 2.1.7 から 2.2

改訂にあたって

- 第 9 回　「数と数式の読み方」として，**2.3**
- 第 10 回　「質問への対処法と国際会議での座長のことば」として，**2.4** と **2.5**
- 第 11 回　「発表用ポスターの作成法と発表法，オンライン発表の心得とオンデマンド発表資料の作成」として，**2.6** と **2.7**
- 第 12 回　「電気磁気学に関係する諸表現」として，**3.1**
- 第 13 回　「電気回路理論に関係する諸表現」として，**3.2**
- 第 14 回　「電気電子工学応用分野に関係する諸表現」として，**3.3** から **3.6** の中の一節(受講者の専門分野に近い節を選択してください)
- 第 15 回　「論文執筆や発表に関係する電子メール文例」として，**付 4**

　最後になりましたが，本書を改訂するにあたり，有益なご助言を賜りました電気学会出版事業委員会の皆様に感謝いたします．また，改訂に際してご協力くださいました電気学会編修出版課の皆様に，心からお礼を申し上げます．

2024 年 10 月吉日

馬場吉弘

目次

第1章　英語論文執筆法　*Techniques for Writing Successful Scientific Papers in English* ― 1

1.1　英語論文の構成　*Structure of a Paper in English* ― 2
- 1.1.1　英語論文の構成　*Structure of a Paper in English*　2
- 1.1.2　題目　*Title*　2
- 1.1.3　著者　*Authors*　5
- 1.1.4　要旨　*Abstract*　6
- 1.1.5　キーワード　*Keywords or Index Terms*　9
- 1.1.6　緒言　*Introduction*　9
- 1.1.7　研究方法　*Method*　12
- 1.1.8　研究結果　*Results*　15
- 1.1.9　考察　*Discussion*　18
- 1.1.10　まとめ　*Summary or Conclusions*　22
- 1.1.11　付録　*Appendices*　23
- 1.1.12　謝辞　*Acknowledgments*　23
- 1.1.13　参考文献　*References*　24
- 1.1.14　著者略歴　*Author Biographies*　27

1.2　能動態と受動態の使い方　*How to Use Active and Passive Voices* ― 27
- 1.2.1　能動態と受動態の関係　*Relation Between Active and Passive Voices*　28
- 1.2.2　能動態の使い方　*How to Use Active Voice*　29

1.3　時制の選び方　*How to Choose Tense* ― 30
- 1.3.1　現在時制の使い方　*How to Use Present Tense*　30
- 1.3.2　未来時制の使い方　*How to Use Future Tense*　32
- 1.3.3　過去時制の使い方　*How to Use Past Tense*　32
- 1.3.4　現在完了形の使い方　*How to Use Present Perfect Tense*　33
- 1.3.5　時制の一致の例外　*Exceptions to the Rule of Sequence of Tenses*　33

1.4　助動詞の使い方　*How to Use Auxiliary Verbs* ― 34
- 1.4.1　"will" と "would" の使い方　*How to Use "will" and "would"*　35
- 1.4.2　"can" と "could" の使い方　*How to Use "can" and "could"*　36
- 1.4.3　"may" と "might" の使い方　*How to Use "may" and "might"*　38
- 1.4.4　"must" の使い方　*How to Use "must"*　39

目次

- 1.4.5 "shall" と "should" の使い方　*How to Use "shall" and "should"*　39
- 1.5 不定冠詞と定冠詞の使い方　*How to Use Indefinite and Definite Articles* ……………41
  - 1.5.1 不定冠詞の使い方　*How to Use Indefinite Articles*　41
  - 1.5.2 定冠詞の使い方　*How to Use Definite Articles*　42
  - 1.5.3 無冠詞の使い方　*How to Use Nouns Without Articles*　44
- 1.6 接続詞, 分詞構文, 関係代名詞の使い方　*How to Use Conjunctions, Participial Constructions, and Relative Pronouns* ……………45
  - 1.6.1 接続詞の使い方　*How to Use Conjunctions*　45
  - 1.6.2 分詞構文の使い方　*How to Use Participial Constructions*　49
  - 1.6.3 関係代名詞の使い方　*How to Use Relative Pronouns*　51
- 1.7 前置詞と群前置詞の使い方　*How to Use Prepositions and Group Prepositions* ……………54
  - 1.7.1 前置詞の使い方　*How to Use Prepositions*　54
  - 1.7.2 群前置詞の使い方　*How to Use Group Prepositions*　59
- 1.8 ハイフン, ダッシュ, コロン, セミコロンの使い方　*How to Use a Hyphen, Dash, Colon, and Semicolon* ……………63
  - 1.8.1 ハイフンの使い方　*How to Use a Hyphen*　64
  - 1.8.2 ダッシュの使い方　*How to Use a Dash*　64
  - 1.8.3 コロンの使い方　*How to Use a Colon*　65
  - 1.8.4 セミコロンの使い方　*How to Use a Semicolon*　67
- 1.9 論文投稿前の文法と剽窃チェックの仕方　*How to Check Grammar and Plagiarism Before Paper Submission* ……………68
  - 1.9.1 文法とつづりのチェックの仕方　*How to Check Grammar and Spelling*　68
  - 1.9.2 剽窃チェックの仕方　*How to Check Plagiarism*　71

第2章　英語論文口頭発表法　*Techniques for Presenting Successful Scientific Papers in English* ——— 73

- 2.1 発表用スライドの作成法と発表法　*How to Prepare Slides for an Oral Presentation and How to Speak with the Slides* ……………74
  - 2.1.1 スライドの構成　*How to Organize Slides*　74
  - 2.1.2 発表の開始　*How to Start a Presentation*　75
  - 2.1.3 発表概要の説明　*How to Show the Outline or Contents*　77
  - 2.1.4 研究の背景と目的の説明　*How to Explain the Background and Objective*　79
  - 2.1.5 研究方法の説明　*How to Explain the Method*　80
  - 2.1.6 結果の説明　*How to Explain the Results*　81
  - 2.1.7 考察の説明　*How to Explain the Discussions*　85
  - 2.1.8 まとめの説明　*How to Explain the Summary or Conclusions*　90

2.2 グラフや表の説明法　*How to Explain Graphs and Tables* ……………………………… 92
　2.2.1 グラフや表の表示　*How to Show a Graph or a Table*　92
　2.2.2 グラフの説明　*How to Explain a Graph*　92
　2.2.3 表の説明　*How to Explain a Table*　94
2.3 数と数式の読み方　*How to Read Numbers and Mathematical Expressions* ……………… 95
　2.3.1 基数と序数　*Cardinal Numbers and Ordinal Numbers*　95
　2.3.2 小数と分数　*Decimals and Fractions*　96
　2.3.3 n 乗根，累乗，対数　*nth Root, Powers, and Logarithms*　97
　2.3.4 足し算と引き算　*Addition and Subtraction*　98
　2.3.5 掛け算と割り算　*Multiplication and Division*　99
　2.3.6 微分と積分　*Differential Calculus and Integral Calculus*　100
　2.3.7 数列，級数，極限　*Sequences, Series, and Limits*　102
　2.3.8 順列，組合せ，確率　*Permutations, Combinations, and Probabilities*　102
　2.3.9 不等式，比較　*Inequalities and Comparison*　103
　2.3.10 複素数　*Complex Numbers*　104
　2.3.11 ベクトル　*Vectors*　105
　2.3.12 行列，行列式　*Matrices and Determinants*　106
2.4 質問への対処法　*How to Handle Questions* ……………………………………………… 106
　2.4.1 質問への応答法　*How to Answer Questions*　106
　2.4.2 コメントへのお礼の仕方　*How to Thank for Comments*　110
　2.4.3 むずかしい質問への対処法　*How to Handle Difficult Questions to Answer*　111
2.5 国際会議での座長のことば　*How to Talk as the Chairperson of a Session at an International Conference* …………………………………………………………………… 112
　2.5.1 セッション前の準備　*How to Prepare for a Session*　112
　2.5.2 セッション開始時のことば　*How to Start a Session*　112
　2.5.3 発表終了時と質疑応答時のことば　*How to Thank a Speaker and Invite Questions*　113
　2.5.4 次の講演に移るときのことば　*How to Move on to the Next Presentation*　115
　2.5.5 セッション締めくくりのことば　*How to Close a Session*　117
2.6 発表用ポスターの作成法と発表法　*How to Prepare a Poster for a Poster Presentation and How to Talk with the Poster* ……………………………………………………………… 118
　2.6.1 ポスターの作成　*How to Prepare a Poster*　118
　2.6.2 概要の説明　*How to Explain the Outline of a Paper*　118
2.7 オンライン発表の心得とオンデマンド発表資料の作成　*Advice about Online Oral Presentation and Preparation of On-demand Oral Presentation Material* ……………… 124
　2.7.1 オンライン発表の心得　*Advice about Online Oral Presentation*　124
　2.7.2 オンデマンド発表資料の作成　*Preparation of On-demand Oral Presentation*

目次

　　　　Material　125

第3章　電気電子工学分野における諸表現　*Various Expressions in Electrical and Electronics Engineering Areas*　127

- 3.1　電気磁気学における表現　*Expressions Used in Electromagnetics* ……… 128
  - 3.1.1　ベクトル解析で用いられる表現　*Expressions Used in Vector Analysis*　128
  - 3.1.2　電界と磁界に関する表現　*Expressions Used for Electric and Magnetic Fields*　132
  - 3.1.3　電磁波に関する表現　*Expressions Used for Electromagnetic Waves*　137
- 3.2　電気回路理論における表現　*Expressions Used in Electrical Circuit Theory* ……… 143
  - 3.2.1　直流回路と交流回路に関する表現　*Expressions Used for Direct-current and Alternating-current Circuits*　143
  - 3.2.2　過渡現象に関する表現　*Expressions Used for Electrical Transient Phenomena*　150
  - 3.2.3　測定器に関する表現　*Expressions for Electrical Measuring Instruments*　155
- 3.3　電力・エネルギー分野における表現　*Expressions Used in Electric Power and Energy Area* ……… 160
  - 3.3.1　電力システムに関する表現　*Expressions for Electric Power Systems*　160
  - 3.3.2　高電圧現象に関する表現　*Expressions for High-voltage Phenomena*　167
  - 3.3.3　エネルギー貯蔵に関する表現　*Expressions for Storage of Energy*　170
- 3.4　電子デバイスに関する表現　*Expressions for Electronics Devices* ……… 175
  - 3.4.1　ダイオードとトランジスタに関する表現　*Expressions for Diodes and Transistors*　175
  - 3.4.2　集積回路に関する表現　*Expressions for Large-scale Integrated Circuits*　181
  - 3.4.3　光電子デバイスに関する表現　*Expressions for Optoelectronics Devices*　184
- 3.5　情報通信に関する表現　*Expressions for Information and Communications* ……… 189
  - 3.5.1　アンテナ・無線通信に関する表現　*Expressions for Antennas and Wireless Communications*　189
  - 3.5.2　光ファイバ通信に関する表現　*Expressions for Optical Fiber Communications*　193
  - 3.5.3　インターネットに関する表現　*Expressions for the Internet*　197
- 3.6　機械学習・人工知能（AI）に関する表現　*Expressions for Machine Learning and Artificial Intelligence* ……… 201
  - 3.6.1　機械学習と深層学習に関する表現　*Expressions for Machine Learning and Deep Learning*　201
  - 3.6.2　機械翻訳に関する表現　*Expressions for Machine Translation*　204
  - 3.6.3　生成AIに関する表現　*Expressions for Generative Artificial Intelligence*　207

目　次

付　録　*Appendices* ─────211

付録 1　論文投稿から掲載または発表にいたるまでの流れ　*General Flow from the Submission of a Scientific Paper to its Publication or Presentation* ………… 212

付 1.1　学術雑誌への論文投稿から掲載にいたるまでの流れ　*General Flow from the Submission of a Paper to a Scientific Journal to its Publication*　212

付 1.2　国際会議での論文発表までの流れ　*General Flow of Procedures Needed for an Oral Presentation at an International Conference*　214

付録 2　インパクトファクタについて　*On Journal Impact Factors* ……………… 216

付録 3　h 指数について　*On the h-index* ……………………………………………… 218

付録 4　電子メール文例　*Examples of Electronic Correspondence (email)* ……… 219

付 4.1　論文査読状況の照会　*Inquiry About the Current Status of a Submitted Paper*　219

付 4.2　著作権で保護された図の使用許可願い　*Request of Permission to Reprint Copyrighted Figures*　220

付 4.3　論文校正刷りの訂正依頼　*Request to Correct Errors in a Proof*　222

付 4.4　雑誌バックナンバーの注文　*Order of a Back Issue*　223

付 4.5　論文送付願い　*Request to Send a Copy of a Paper*　224

付 4.6　ビザ（査証）取得のための招待状送付願い　*Request of an Invitation Letter for Acquiring a Visa*　226

付 4.7　博士研究員としての留学の打診　*Application for a Postdoctoral Position*　227

付録 5　アメリカ留学関連手続き　*Procedures Needed for Studying in the United States* ………… 233

付 5.1　日本出国前の準備　*Preparation in Japan Before Leaving for the United States*　233

付 5.2　アメリカ入国後の手続き　*Procedures Needed Right After Arrival in the United States*　236

付録 6　電気学会誌に掲載された英語論文執筆，発表法に関する文献リスト　*List of IEEJ Journal Articles on Writing and Presenting Scientific Papers in English* …… 238

参　考　文　献 ───── 240
索　　　　引 ───── 242
著者および監修者紹介 ───── 262

第 1 章

英語論文執筆法

Techniques for Writing Successful Scientific Papers in English

1.1 英語論文の構成
1.2 能動態と受動態の使い方
1.3 時制の選び方
1.4 助動詞の使い方
1.5 不定冠詞と定冠詞の使い方
1.6 接続詞，分詞構文，関係代名詞の使い方
1.7 前置詞と群前置詞の使い方
1.8 ハイフン，ダッシュ，コロン，セミコロンの使い方
1.9 論文投稿前の文法と剽窃チェックの仕方

1.1 英語論文の構成
Structure of a Paper in English

本節では，学術雑誌(academic or scientific journals)や国際会議(international scientific conferences)に投稿する英語論文の基本的な構成について説明します．

1.1.1 英語論文の構成　*Structure of a Paper in English*

学術雑誌や国際会議には，論文題目の文字サイズや文字数，要旨の語数，キーワードの語数，参考文献の表記法などの詳細について記した独自の投稿規定(Guide for authors または Instructions to authors)があります．多くの場合，投稿規定とともに MS-Word 形式の論文テンプレートファイルも提供されます．研究成果の投稿先を決めたら，その学術雑誌あるいは国際会議のウェブページから投稿規定を入手し，その規定に従って論文を書く必要があります．なお，投稿規定に，"The paper text must be double spaced." と記されている場合がありますが，"double space" とは行と行の間に 1 行分に相当する空白をあけるという意味です(A4 サイズ 1 ページあたりで約 25 行になります)．

上述のように，学術雑誌や国際会議により細部に相違はありますが，基本的な英語論文の構成は**表 1.1**のようになります．

1.1.2 題目　*Title*

学術雑誌や国際会議のプロシーディングス(proceedings)に掲載されている論文を読むか否かを，読者はまず題目で判断します．したがって，どのような内容の論文であるかをわかるように，内容を表す重要なキーワードを含めて，短すぎず長すぎない題目をつける必要があります(100 字程度以内で)．このように題目を設定すると，読者による文献データベースでの論文検索にも役立ちます．あまりよく知られていない専門用語(jargon)の使用は避けるのが無難です．また，"laser (light amplification by stimulated emission of radiation)" のようによく知られているもの以外の頭字語(acronym)や短縮形(abbreviation．たとえば，equation を eq.と短縮)の使用も避けるのが無難です．題目が長くなるとしても，もとの熟語を使用しましょう．一方，"A

表 1.1　英語論文の基本構成

順番	項　目
1	題　　目　（Title）
2	著　　者　（Authors）
3	要　　旨　（Abstract）
4	キーワード　（Keywords or Index Terms）
5	緒　　言　（Introduction）
6	研究方法　（Method）
7	研究結果　（Results）
8	考　　察　（Discussion）
9	まとめ　（Summary or Conclusions）
10	付　　録　（Appendices）
11	謝　　辞　（Acknowledgments）
12	参考文献　（References）
13	著者略歴　（Author Biographies）

report on …"（…についての報告）や "A study of …"（…の研究）などの語句は英語論文では思い切って省略してもよいでしょう．

「電力線搬送通信信号の伝搬に与える電力ケーブルの半導電層の影響を明らかにする目的で，時間領域有限差分法を用いて数値シミュレーションを行い，それにより得られた結果について考察した」論文を書く場合を例に，その題目を考えてみましょう．この論文の重要なキーワードが，順に「半導電層の影響」(influence of semiconducting layer)，「電力ケーブル (power cable)」，「電力線搬送通信信号 (power line communication signal)」，「伝搬 (propagation)」，「時間領域有限差分法 (finite-difference time-domain method)，または時間領域有限差分解析 (finite-difference time-domain analysis)」の五つであるとし，それらをすべて含めると下記のような題目が考えられます．なお，英文題目の下には日本語訳を示しています．

Finite-difference time-domain analysis of the influence of semiconducting layers on propagation of power line communication signals in power cables

電力ケーブル上での電力線搬送通信信号の伝搬に与える半導電層の影響についての時間領域有限差分解析

この題目の語間の空白を含めた文字数は147ですので,少し長すぎます."finite-difference time-domain" および "power line communication" の頭字語である "FDTD" および "PLC" が,投稿する学術雑誌の読者や国際会議の参加者の多くに知られている場合には,それらを用いて下記のように表記できます.この場合の空白を含めた字数は101で,多くの学術雑誌や国際会議において許容される長さです.

> FDTD analysis of the influence of semiconducting layers on propagation of PLC signals in power cables

"FDTD" および "PLC" という頭字語があまり知られていない場合には,優先順位の低いキーワードの一つ "finite-difference time-domain analysis" を外して,たとえば,次のような題目にできます.

> Influence of semiconducting layers on propagation of power line communication signals in power cables
>
> 電力ケーブル上での電力線搬送通信信号の伝搬に与える半導電層の影響

この場合も,空白を含めた字数は101で,多くの学術雑誌や国際会議において許容される長さです.ただし,"finite-difference time-domain analysis" というキーワードを外していますので,どのような手法によって,この研究を行ったかは,題目のみからではわからなくなります.

検討対象の電力ケーブルの具体的な電圧レベルを含めたい場合には,次のような題目にできます.空白を含めた字数は101となり,ちょうどよい長さです.

> Influence of semiconducting layers on propagation of power line communication signals in 66-kV cables
>
> 66 kV ケーブルにおける電力線搬送通信信号の伝搬に与える半導電層の影響

題目は,上記のように第1語の頭文字のみを大文字(capital または upper case)で表記する場合(ただし,固有名詞などは第1語ではなくても頭文字を大文字にします)のほか,下記のように内容語(名詞,動詞,形容詞,副詞な

ど)の頭文字と機能語(冠詞,前置詞,接続詞)のうちの長い(5字以上程度の)前置詞と接続詞の頭文字を大文字で表記する場合,および題目のすべてを大文字で表記する場合があります.

> Influence of Semiconducting Layers on Propagation of Power Line Communication Signals in Power Cables
>
> INFLUENCE OF SEMICONDUCTING LAYERS ON PROPAGATION OF POWER LINE COMMUNICATION SIGNALS IN POWER CABLES

なお,理工学系の学術論文の題目は,上記のような名詞句構造のものがほとんどですが,下記のような前置詞で始まる題目や文構造の題目もときどき見られます.

> On the mechanism of X-ray production by dart leaders of lightning flashes
>
> 雷の前駆放電によるエックス線発生のメカニズムについて
>
> Does Wilson's cloud chamber provide clues on lightning initiation in thunderclouds?
>
> ウィルソンの霧箱(実験)は雷雲内の雷放電発生についての手掛かりを与えるか?

1.1.3 著者 *Authors*

研究の本質的な部分に携わった方々の名は,その方々の同意を得たうえで,著者に含めます.著者の順は,その研究や論文への貢献の度合いの高い順とするのが基本です.ちょっとしたお手伝いをしてくれた方,助言をくれた方,議論してくれた方については,論文の謝辞の部分で謝意を述べることができます.最近は,不適切なオーサーシップ(inappropriate authorship)が研究活動上の不正行為として広く認識されるようになりました.これは,その論文への貢献があまりない方が著者に加えられたり,貢献が大きい方が著者に加えられなかったりすることなどを指します.この問題を防ぐために,最近

は，論文投稿時に，各著者が具体的にどのような貢献を行ったかを記載することや，著者としての条件を満たしているか否かを確認する質問群に回答することが求められます．その論文が高い評価を得た場合に，各著者は名誉を得ることができる一方で，論文に大きな誤りや不正が見つかった場合などには，著者本人や各著者の所属機関の信用までも低下させてしまうなど，思いがけない災難に巻き込んでしまう可能性もあります．このため，著者に加えるべきか否かや，著者の順については，指導教授や上司と慎重に相談して決めましょう．

たとえば，大学院生の電気花子さんと電子太郎教授の共同で行った研究で，その研究および論文への電気花子さんの貢献度が高い場合，その論文の著者欄には，次のように表記します．

Hanako Denki,　and　Taro Denshi

電気　花子　　　　電子　太郎

なお，"Mr." や "Dr." などの称号はつけませんが，その雑誌を発行する学会の会員種別を下記のように付記する場合もあります．

Hanako Denki, Student Member,　and　Taro Denshi, Senior Member

学生員　電気　花子　　　　　　上級会員　電子　太郎

1.1.4　要旨　*Abstract*

題目を読んでその論文に興味をもった読者は，次に要旨を読みます．要旨を読んで，その論文本体を読み進めるか否かを判断します．したがって，その論文のなかで最も重要な内容を記述します．要旨は，それ自体で完結している必要があり，どのような問題を対象とし，どのような解析や実験を行い，どのような結果を得て，どのようなことを明らかにしたかを1段落(paragraph)で記述します．要旨も読者による文献データベースでの検索対象になります．

要旨では，通常，参考文献は引用しませんし，図や式も使用しません．頭字語については，初出時にもとの熟語と一緒に示せば，2回目以降は頭字語のみで使用できます．最近では，要旨であっても，著者を主語(we, the authors など)とした能動態の文も用いられており,受動態で記述しなければならないわけではありません.時制については,読者がその論文を読んでいる瞬間を現在と考え,それを基準にして,現在形のほか,過去形や現在完了形などを用います.能動態と受動態の選び方については 1.2 で,時制の選び方については 1.3 で詳しく述べます.

学術雑誌や国際会議にもよりますが,要旨の長さは 100〜200 語(words)程度が一般的です.下記は,「電力ケーブル上での電力線搬送通信信号の伝搬に与える半導電層の影響」という題目の論文の要旨の例です.語数は 193 です.

Power-line-communication (PLC) systems use high-voltage distribution lines and cables for data communications in a frequency range up to 30 MHz. Since the distribution lines and cables are not designed for effectively transmitting the PLC signals, they might attenuate and distort significantly along the lines and cables. We have computed the propagation characteristics of a PLC signal of frequency 30 MHz along a single-core power cable having two, 3-mm thick semiconducting layers, using the finite-difference time-domain (FDTD) method. We found from the FDTD computations that the PLC signal suffers significant attenuation particularly when the semiconducting-layer conductivity σ is about 10^{-3} and 10^3 S/m. The reason for the significant signal attenuation around $\sigma=10^{-3}$ S/m is that charging and discharging processes in a radial direction of the semiconducting layers become marked since the time constant (27 ns) of the semiconducting layer is close to the half cycle (17 ns) of a 30-MHz signal. The reason for the significant signal attenuation around $\sigma=10^3$ S/m is that axial current flows in the semiconducting layers because a penetration depth (2 mm) for frequency 30 MHz and $\sigma=10^3$ S/m is close to the semiconducting-layer thickness (3 mm).

電力線搬送通信(PLC)システムでは,30 MHz にいたるまでの周波数のデータ通信に電力線や電力ケーブルが使用されます.電力線や電力ケーブルは PLC 信号を効果的に伝送するようには設計されてはいないため,これらは電力線や電力ケーブル上で著しく減衰および歪みする可能性があります.私たちは,時間領域有限差分(FDTD)法を用いて,厚み 3 mm の半導電層を 2 層有する単芯電力ケーブル上での周波数 30

MHz の PLC 信号の伝搬特性の計算を行いました．この FDTD 計算により，半導電層の導電率 σ が 10^{-3} および 10^3 S/m 程度のときに，PLC 信号は著しく減衰することを発見しました．$\sigma=10^{-3}$ S/m の場合に生じる信号の著しい減衰は，半導電層の時定数(27 ns) が 30 MHz 信号の半サイクル(17 ns)に近くなることによる半導電層の半径方向の充放電現象が原因です．$\sigma=10^3$ S/m の場合に生じる信号の著しい減衰は，周波数 30 MHz，導電率 $\sigma=10^3$ S/m に対する透過深さ(2 mm)が半導電層の厚さ(3 mm)に近くなることにより半導電層に軸方向電流が流れることが原因です．

この要旨の "We found from the FDTD computations that a 30-MHz PLC signal suffers significant attenuation..." では，主節(main clause)の動詞が過去形(found)，従属節(subordinate clause)の動詞が現在形(suffers)となっており，時制が一致していません．この文には，FDTD 計算を実施し従属節で述べられていることを発見したのは，読者の「現在」を基準にすると「過去」のことですが，従属節で述べられている「半導電層の導電率 σ が 10^{-3} および 10^3 S/m 程度のときに，PLC 信号は著しく減衰する」ことは単なる計算結果ではなく，再現性のある事実であるという意図が込められています(時制の一致の例外)．時制については 1.3 で詳しく説明します．

雑誌によっては，要旨の長さが 100 語に制限されていたり，要旨の最初の 100 語しかウェブに示されない場合もあります．上記の 193 語の要旨をさらに要約し，99 語にしたものを以下に示します．

Power-line-communication (PLC) systems use high-voltage cables in a frequency range up to 30 MHz. Since the cables are not designed for this duty, there can be significant signal attenuation and distortion. We have computed the propagation characteristics of a typical PLC signal along a single-core cable having two, 3-mm thick semiconducting layers, using the finite-difference time-domain (FDTD) method. The PLC signal suffers significant attenuation for two different values of layer conductivity σ. With $\sigma=10^{-3}$ S/m, attenuation is caused by radial charging and discharging of the semiconducting layers. Axial current propagation along the layers dominates losses at $\sigma=10^3$ S/m.

電力線搬送通信(PLC)システムは，30 MHz にいたるまでの周波数範囲で高電圧ケーブルを利用します．高電圧ケーブルはこのような責務の

ために設計されてはいませんので，信号は著しく減衰および変わいする可能性があります．私たちは，時間領域有限差分(FDTD)法を用いて，厚み3mmの半導電層を2層有する単芯電力ケーブル上での標準的なPLC信号の伝搬特性の計算を行いました．半導電層の導電率が異なる二つの場合に，PLC信号は著しく減衰します．$\sigma=10^{-3}$ S/m の場合，減衰は半導電層の半径方向の充放電現象により生じます．$\sigma=10^{3}$ S/m の場合の損失は，半導電層の軸方向電流伝搬が原因です．

1.1.5　キーワード　*Keywords or Index Terms*

読者が，文献データベースで興味のある論文を探す際に役立つキーワードを数語記しておきましょう．ある程度自由にキーワードを設定できる雑誌もあれば，雑誌が指定するキーワードリストから選定しなければならない場合もあります．以下では，五つのキーワードを示しています．

Semiconducting layer, power cable, power line communication (PLC) signal, propagation, finite-difference time-domain (FDTD) method.

半導電層，電力ケーブル，電力線搬送通信(PLC)信号，伝搬，時間領域有限差分(FDTD)法．

1.1.6　緒言　*Introduction*

緒言では，最初に，研究の背景について述べます．次に，関連する先行研究を，それらの研究に対応する文献を引用しながら紹介します．そのうえで，なぜ，この研究を行う必要があったのか，どのような重要な課題が残されているのかを示します．最後に，この研究では，どのような問題を対象とし，どのような解析や実験を行い，どのようなことを明らかにしようとしているのか(研究の目的)を述べます．

なお，禁止されているわけではありませんが，緒言では図はあまり用いられません．頭字語については，初出時にもとの熟語と一緒に示せば，2回目以降は頭字語のみで使用できます．ただし，要旨(abstract)は独立したものと考えますので，要旨において頭字語をもとの熟語と一緒に示していたとしても，緒言における初出時に改めてもとの熟語とともに示す必要があります．

第 1 章　英語論文執筆法

　下記は,「電力ケーブル上での電力線搬送通信信号の伝搬に与える半導電層の影響」という題目の論文の緒言の例です．第 1 段落でこの研究の背景が,第 2 段落で先行研究の紹介が,第 3 段落でこの研究の必要性が,第 4 段落でこの研究の目的が述べられています．

　Recently, considerable attention has been attracted to the power line communication (PLC). The PLC systems use power distribution lines and cables for data communications in a frequency range up to 30 MHz. Within a power cable, semiconducting layers are usually incorporated between the core conductor of the power cable and the insulating layer, and between the insulating layer and the sheath conductor. These layers mitigate electrical breakdown from unexpected projections on the core and sheath conductor surfaces, and also accommodate the thermal expansion of the insulating layer. Since power cables are not designed for effectively transmitting the PLC signals, they might attenuate significantly along the cables owing to the presence of semiconducting layers [1][2].

　Steinbrich [3] has shown, experimentally and computationally with a distributed-circuit representation for a single-core power cable having semiconducting layers, that PLC signals of a frequency range up to 30 MHz attenuate significantly (about 4 dB/100 m at 30 MHz) along a 20-kV crosslinked polyethylene insulated (XLPE) cable. This XLPE cable included semiconducting carbon-polyethylene compound layers of conductivity $\sigma=17$ S/m and semiconducting carbon-filled paper layers of conductivity $\sigma=0.05$ S/m. On the other hand, Cataliotti et al. [4] have shown experimentally that the attenuation of PLC signals of a frequency range from 25 kHz to 0.2 MHz along a power cable with semiconducting layers is less significant (about 1 dB/km at 0.2 MHz), and the measured attenuation constant variation within this frequency range is reasonably well reproduced by their simplified equivalent circuit representation that neglects the presence of the semiconducting layers.

　It appears from these results that power cables with semiconducting layers cause significant attenuation only for signals of frequencies above 1 MHz. However, the quantitative relation between the attenuation of PLC signals and the conductivity of power-cable semiconducting layers has not yet been clarified, and qualitative mechanisms for signal attenuation have not been fully understood.

The objective of this study is to reveal the influences of the semiconducting layers on the PLC propagation characteristics. For this objective, we compute the propagation characteristics of a PLC signal of frequency 30 MHz along a single-core power cable having two, 3-mm thick semiconducting layers, using the finite-difference time-domain (FDTD) method [5] in the two-dimensional cylindrical coordinate system.

　最近，電力線搬送通信（PLC）が大変注目されています．PLC システムでは，30 MHz にいたるまでの周波数のデータ通信に電力線や電力ケーブルが使用されます．電力ケーブルには，その中心導体と絶縁層および絶縁層とシース導体の間に半導電層が通常設けられています．これらの半導電層は，中心導体表面やシース導体表面における予想外の突起から絶縁破壊が発生するのを抑制し，また絶縁層の熱膨張に順応します．電力ケーブルは，PLC 信号を効率的に伝送するようには設計されていませんので，半導電層の存在により，PLC 信号は電力ケーブル上で著しく減衰する可能性があります[1][2]．

　スタインリッヒ[3]は，実験的に，そして半導電層を有する単芯電力ケーブルの分布定数回路モデルを用いた計算により，20 kV 用の架橋ポリエチレン絶縁（XLPE）ケーブルにおいて，30 MHz までの PLC 信号が著しく減衰することを示しています（30 MHz 信号の減衰は約 4 dB/100 m）．この XLPE ケーブルは，導電率 17 S/m の炭素ポリエチレン化合物の半導電層と導電率 0.05 S/m の炭素含有紙の半導電層を含んでいました．一方，カタリオッティら[4]は，半導電層を含む電力ケーブルにおいて，25 kHz から 0.2 MHz までの PLC 信号の減衰はそれほど顕著ではなく（0.2 MHz 信号の減衰は約 1 dB/km），半導電層の存在を無視した簡易等価回路により，この周波数範囲で実測した減衰定数の変化を良好に再現できることを示しています．

　これらの結果より，半導電層を含む電力ケーブルは 1 MHz 以上の周波数の信号のみを著しく減衰させることがわかります．しかし，PLC 信号の減衰と電力ケーブルの半導電層の導電率との定量的な関係はまだ明らかにされていません，また，信号の減衰の定性的なメカニズムも完全に理解されるまでにはいたっていません．

　本研究の目的は，PLC 信号伝搬特性に与える半導電層の影響を明ら

> かにすることです．この目的のため，私たちは，二次元円筒座標系における時間領域有限差分(FDTD)法[5]を用いて，厚み3mmの半導電層を2層有する単芯電力ケーブル上での周波数30MHzのPLC信号の伝搬特性の計算を行います．

[1]〜[5]は参考文献番号です．参考文献の引用箇所での表示法や参考文献の列挙の方法については，1.1.13で詳しく説明します．

第2段落で，スタインリッヒとカタリオッティらの先行研究の紹介をしています．どちらも過去のことですが，それらの結果が現在のこの論文にも影響しているという意識を込めて，現在完了形を用いています．時制については 1.3 で詳しく説明します．"from 25 kHz to 0.2 MHz" のような "from A to B" は，通常は範囲として起点Aと着点Bを含みます．このことを確実に表したい場合，"from A to B inclusive" と書きます．一方，AとBを含まない場合には，"from A to B exclusive" と書きます．

「本研究の目的は…です．」は "The purpose of this study is …"，"The aim of this study is …"，"The goal of this study is …" と書くこともできます．8ページを超えるような長い論文の場合には，研究の目的を述べた後，次のように論文の構成の説明を付け加えましょう．"This paper is organized as follows（または The structure of this paper is as follows）. In Section 2, we present … In Section 3, we show …"．

1.1.7　研究方法　*Method*

研究方法では，採用した実験方法，計算方法あるいは計算モデルについて説明します．この部分を読んだほかの研究者が，その実験や計算を再現する際に必要な具体的な情報を記述します．

下記は，「電力ケーブル上での電力線搬送通信信号の伝搬に与える半導電層の影響」という題目の論文における計算モデルの説明例です．使用した計算法，計算モデルの詳細，計算時間刻み幅，空間刻み幅，使用したプログラミング言語，使用した計算機の仕様，計算時間が述べられています．なお，使用した計算法(FDTD法)は，この論文の投稿先となる雑誌の読者の多くに知られている可能性が高いため，緒言においてその計算法について述べられた文献を引用するに留め，計算法自体についての説明はしていません．しかし，読者にあまり知られていない計算法を用いた場合には，その計算法自体についても説明する必要があります．実験の場合には，使用した実験装置や器具の仕様，実験対象の諸元について詳しく説明する必要があります．

なお，読者がその論文を読んでいる瞬間を現在と考えると，実験が行われたのは過去のことになりますので，実験の説明には過去形が用いられることが多いようです．一方，計算方法や計算モデルについては，読者は，計算式やモデル図を見ながら，対応する計算方法や計算モデルの説明を読みますので，それらの説明には現在形が用いられることが多いようです．

> Figure 1 shows a 130-m long single-core power cable, to be analyzed using the FDTD method. The radius of the core conductor is 5 mm, and the inner radius of the sheath conductor is 25 mm. Both the core and sheath conductors are perfectly conducting. Figure 1 also shows a 14-mm thick insulating layer with semiconducting layers of 3-mm thickness on both inner and outer surfaces. The relative permittivity ε_r of the insulating layer and of each semiconducting layer is set to $\varepsilon_r=3$. The conductivity of each semiconducting layer is set to a value in a range from $\sigma=10^{-5}$ to 10^5 S/m. Note that this power cable is rotationally symmetric around its axis and has a circular cross-section. It is represented without staircase-approximated contour in the FDTD method using a 2D cylindrical coordinate system with a working space of 130 m×27 mm rectangles, contoured by the thick black line shown in
>
>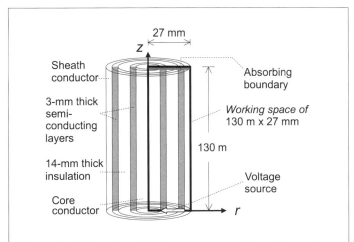
>
> Figure 1. 130-m long single-core power cable with semiconducting layers, to be analyzed using the FDTD method. The 130 m×27 mm rectangle space, contoured by a thick black line, is the actual working space for the present FDTD computations in the 2D cylindrical coordinate system.

図1.1　計算モデル図の例

Figure 1.

At one of the ends of the cable, a 10-V positive half-sine pulse of frequency $f=30$ MHz is applied between the core and sheath conductors. The other end of the cable model is terminated using Liao's second-order absorbing boundary [6]. The working space of 130 m × 27 mm for the FDTD computation is divided into 1 mm × 1 mm square cells. The time increment is set to $\Delta t=2.3$ ps, which is 97.5 % of the upper limit of Courant's stability condition that $\Delta t \leq \Delta r/c/\sqrt{2}$, where Δr is the side length of square cell employed and c is the speed of light. Computations are carried out up to 900 ns with the FDTD computation program, which is programmed by the authors in C++ on the basis of the standard second-order accurate FDTD equations. The computation time is about 20 hours when a personal computer with a 3-GHz processor and an 8-GB memory is used.

　図1は，FDTD法を用いた解析に使用する長さ130 mの単芯電力ケーブルを示しています．中心導体の半径は5 mm，シース導体の内側半径は25 mmです．中心導体およびシース導体はともに完全導体です．図1は，内側および外側表面の両方に厚さ3 mmの半導電層をもつ厚さ14 mmの絶縁層も示しています．絶縁層および各半導電層の比誘電率は$\varepsilon_r=3$に設定されています．半導電層の導電率は$\sigma=10^{-5}\sim10^5$ S/mの範囲の値に設定されています．なお，この電力ケーブルは，その軸を中心として回転対称で，円形の断面を有しています．この電力ケーブルは，二次元円筒座標系のFDTD法により，階段近似せずに表現されています．図1の太い黒の実線で囲まれた130 m×27 mmの長方形平面が計算領域です．

　このケーブルの一端の中心導体とシース導体間に，振幅10 V，周波数$f=30$ MHzの正弦波の単一正半波が印加されています．このケーブルの他端はリャオの二次吸収境界[6]で終端されています．130 m×27 mmのFDTD計算領域は，1 mm×1 mmの正方形セルに分割されています．計算時間刻み幅は$\Delta t=2.3$ psに設定されています．この値はクーラン安定条件の上限である$\Delta t\leq\Delta r/c/\sqrt{2}$の97.5%です．ここで，$\Delta r$は正方形セルの一辺の長さ，$c$は光速です．計算は，二次精度のFDTD計算式に基づき，C++で著者らにより作成されたFDTD計算プログラムを用いて行われています．3 GHzのプロセッサと8 GBのメモ

リをもつパソコンを用いた場合の計算時間は約20時間です．

図の下側（表の場合は上側）に記述する説明文を"legend"または"caption"といいます（"caption"は図の下側の説明文の他，図中の説明，上記の例では"Sheath conductor"や"Voltage source"なども含みます）．本文において，各図表についての詳細な説明をしたとしても，図または表の"legend"を読んだだけで，それがなにを表しているかを理解できるように，ある程度詳しく記述しましょう．なお，図表の"legend"における最初の名詞（上記の例では130-m long single-core power cable）の冠詞（a，an あるいは the）は通常省略されます．

1.1.8 研究結果　*Results*

研究結果では，実験や計算により得られた結果をグラフ化して示します．表よりもグラフのほうが短時間で情報を伝えられるため，できるだけグラフにして示しましょう．

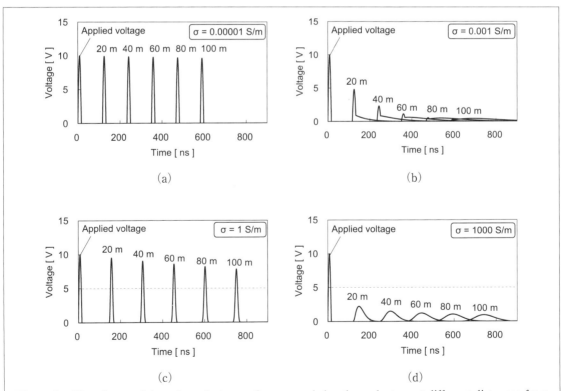

Figure 2. Waveforms of the voltage between the core and sheath conductors at different distances from the excitation point when a 30-MHz and 10-V half-sine pulse is injected and the semiconducting-layer conductivity σ is (a) 10^{-5}, (b) 10^{-3}, (c) 1, and (d) 10^3 S/m.

図1.2　グラフ化した計算結果の例1

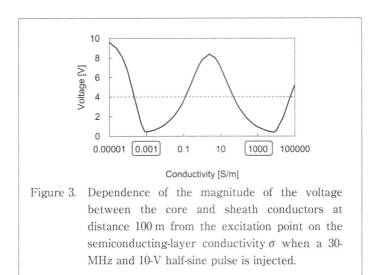

Figure 3. Dependence of the magnitude of the voltage between the core and sheath conductors at distance 100 m from the excitation point on the semiconducting-layer conductivity σ when a 30-MHz and 10-V half-sine pulse is injected.

図 1.3　グラフ化した計算結果の例 2

下記は，「電力ケーブル上での電力線搬送通信信号の伝搬に与える半導電層の影響」という題目の論文における研究結果の説明例です．

Figures 2 (a), (b), (c), and (d) show waveforms of the voltage between the core and sheath conductors of the power cable at different distances of 20, 40, 60, 80 and 100 m from the excitation point when the frequency is $f=30$ MHz and 10-V pulse is injected. The semiconducting-layer conductivity is set to $\sigma=10^{-5}$, 10^{-3}, 1, and 10^3 S/m for computing Figures 2 (a), (b), (c), and (d), respectively. Figure 3 shows how σ affects the magnitude of the voltage between the core and sheath conductors at distance 100 m from the excitation point. Figure 4 shows the dependence of the propagation velocity of a 30-MHz pulse on σ. Note that the propagation velocity is calculated from the propagation distance of 100 m from the voltage excitation point and the corresponding propagation time, evaluated by tracking the voltage pulse peak.

It follows from Figures 2 and 3 that the magnitude of a voltage pulse decreases with increasing propagation distance in all cases considered. However, the dependence of the signal attenuation on σ is not monotonic: the attenuation is significant around $\sigma=10^{-3}$ and 10^3 S/m, while it is not when σ is lower than about 10^{-5} S/m or σ is around 1 S/m. When $\sigma=10^{-3}$ S/m and 10^3 S/m, dispersion is also marked.

It follows from Figure 4 that the propagation velocity of a 30-MHz half-

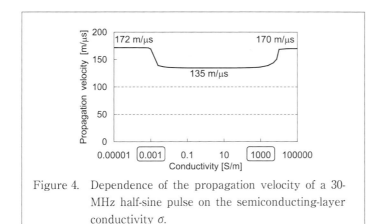

Figure 4. Dependence of the propagation velocity of a 30-MHz half-sine pulse on the semiconducting-layer conductivity σ.

図 1.4 グラフ化した計算結果の例 3

sine pulse is about 135 m/μs when $10^{-2} \leq \sigma \leq 10^{2}$ S/m, while it is about 170 m/μs when $\sigma \leq 10^{-4}$ S/m or $\sigma \geq 10^{4}$ S/m.

　図 2(a), (b), (c)および(d)は，周波数が $f=30$ MHz で 10 V のパルスが印加された場合の電圧印加点から 20, 40, 60, 80 および 100 m 離れた点における電力ケーブルの中心導体とシース導体間の電圧波形です．図 2(a), (b), (c)および(d)の計算に用いた半導電層の導電率は，$\sigma=10^{-5}$, 10^{-3}, 1 および 10^{3} S/m です．図 3 は，電圧印加点から 100 m 離れた点での中心導体とシース導体間電圧振幅に，半導電層導電率 σ がどのような影響を与えるのかを示しています．図 4 は，30 MHz のパルス伝搬速度の導電率 σ への依存性を示しています．なお，伝搬速度は，電圧印加点からの 100 m の伝搬距離と，電圧ピーク値部分で評価した対応する伝搬時間から算出しています．

　図 2 および図 3 より，考慮したすべての場合において，電圧パルスの振幅は伝搬とともに減衰することがわかります．しかし，減衰の半導電層導電率 σ への依存性は単調ではありません．$\sigma=10^{-5}$ S/m 以下および 1 S/m 付近において減衰は顕著ではない一方で，$\sigma=10^{-3}$ および 10^{3} S/m 付近では減衰が顕著となっています．$\sigma=10^{-3}$ および 10^{3} S/m には，分散も顕著になっています．

　図 4 より，30 MHz の正弦波半パルスの伝搬速度は，$\sigma \leq 10^{-4}$ S/m または $\sigma \geq 10^{4}$ S/m の場合には約 170 m/μs ですが，$10^{-2} \leq \sigma \leq 10^{2}$ S/m の場合には約 135 m/μs となっています．

伝搬あるいは進む方向も意識した（ベクトル的な）「速度」は "velocity" です．伝搬方向などを意識しない絶対値的な「速さ」には "speed" を使用します．

パラメータ値のみが異なる同種類の複数のグラフは，Figure 2 のように "legend" を共有して示すと（compound figure），結果を比較するうえで便利です．コロン（：）は，コロンの前の内容を詳しく説明したり，具体例をあげたり，細目を列挙したりする場合に用いられます．コロン（：）やセミコロン（；）の使い方については 1.8 で詳しく説明します．

1.1.9　考察　*Discussion*

考察では，研究結果で示した新しい実験結果や計算結果に物理的解釈を与え，どのような意味があるのかについて説明します．

下記は，「電力ケーブル上での電力線搬送通信信号の伝搬に与える半導電層の影響」という題目の論文における考察の例です．節を二つに区分し，電力線搬送通信信号の減衰と半導電層の導電率との関係，そして信号伝搬速度と半導電層の導電率との関係を説明しています．

A. Attenuation

We discuss here the significant attenuation of a 30-MHz signal propagating along a power cable when the semiconducting-layer conductivity is about $\sigma=10^{-3}$ and 10^3 S/m as shown in Figures 2 and 3. The signal attenuation around $\sigma=10^{-3}$ S/m is probably caused by capacitive charging and discharging of the semiconducting layers in radial direction. For $\sigma=10^3$ S/m, axial current propagation in the semiconducting layers is the dominant cause of attenuation. We quantify these below.

The time constant τ of each semiconducting layer is given by

$$\tau = CR = \frac{2\pi\varepsilon_0\varepsilon_r}{\ln(r_2/r_1)} \frac{\ln(r_2/r_1)}{2\pi\sigma} = \frac{\varepsilon_0\varepsilon_r}{\sigma}, \tag{1}$$

where $C=2\pi\varepsilon_r\varepsilon_0/\ln(r_2/r_1)$ is the per-unit-length capacitance of each semiconducting layer, $R=\ln(r_2/r_1)/2\pi\sigma$ is its radial-direction per-unit-length resistance, r_2 is the outer radius of the semiconducting layer, r_1 is its inner radius, ε_0 is the permittivity of vacuum, ε_r is the relative permittivity of the semiconducting layer, and σ is its conductivity. For $\varepsilon_r=3$ and $\sigma=10^{-3}$ S/m, the time constant is $\tau=27$ ns. Charging and discharging processes in the radial direction of the semiconducting layer with $\sigma=10^{-3}$ S/m will have a strong effect on a 30-MHz signal with 17-ns half cycle, leading to significant attenuation and distortion. Note that, in this condition, the magnitude of the radial conduction current across the semiconducting layer is close to that of the radial displacement current. In other words, the conductance of the semiconducting layer is close to its susceptance.

At high conductivity of $\sigma=10^3$ S/m, the depth d of penetration for an electromagnetic wave of frequency f into a medium of conductivity σ and permeability μ_0 is relevant for loss calculations. This depth is given by

$$d=1/\sqrt{2\pi f \sigma \mu_0}. \tag{2}$$

For $\mu_0=4\pi\times10^{-7}$ H/m, $f=30$ MHz, and $\sigma=10^3$ S/m, Equation (2) yields $d=2$ mm, which is close to the thickness of the semiconducting layer (3 mm). Therefore, most of the axial current flows in the semiconducting layers rather than on the core and sheath conductor surfaces. This results in the significant signal anntenuation and dispersion.

B. Propagation Velocity

We discuss here propagation velocity of a 30-MHz signal along a power cable. Figure 4 shows that the propagation velocity is about 170 m/μs when $\sigma\leq10^{-4}$ S/m, acting as insulators, or $\sigma\geq10^4$ S/m, when the semiconducting layers act as conductors. Therefore, the signal propagation velocity is theoretically determined by the relative permittivity values ε_r of the insulating and semiconducting layers or that of the insulating layer, respectively, which is given by $300/\sqrt{\varepsilon_r}=173$ m/μs.

When σ ranges from 10^{-2} to 10^2 S/m, the propagation velocity is about 135 m/μs, which is lower than the propagation velocities for $\sigma\leq10^{-4}$ S/m and $\sigma\geq10^4$ S/m. For $\sigma=1$ S/m, centered in the range from 10^{-2} to 10^2 S/m, Equation (2) yields $d=65$ mm for $f=30$ MHz. Since this penetration depth is larger than the thickness of each semiconducting layer (3 mm), the axial current propagates mainly on the surface of the core conductor and the inner surface of the sheath conductor, and the magnetic field energy is stored between the surface of the core conductor and the inner surface of the sheath conductor. The corresponding inductance is, therefore, given by $L'=\mu_0 \ln(25\text{ mm}/5\text{ mm})/2\pi$. Since the time constant of the 1-S/m semiconducting layer is $\tau=0.027$ ns from Equation (1), charge would move radially from the surface of the core conductor to the outer surface of the inner semiconducting layer, and from the inner surface of the sheath conductor to the inner surface of the outer semiconducting layer immediately. Therefore, electric field energy is stored in the insulating layer, and the corresponding capacitance is given by $C'=2\pi\varepsilon_0\varepsilon_r/\ln[(25\text{ mm}-3\text{ mm})/(5\text{ mm}+3\text{ mm})]$. From L' and C', the corresponding signal propagation velocity is theoretically given by

$$v' = \frac{1}{\sqrt{L'C'}} = \sqrt{\frac{2\pi}{\mu_0 \ln(25/5)} \frac{\ln(22/8)}{2\pi\varepsilon_0\varepsilon_r}} = 132 \text{ m}/\mu\text{s}, \quad (3)$$

which agrees well with the FDTD-computed propagation velocity (135 m/μs).

A. 減衰

ここでは，図2および図3に示されている半導電層の導電率が $\sigma=10^{-3}$ および 10^3 S/m 程度の場合における電力ケーブル上での30 MHz 信号の著しい減衰について考察します．半導電層の導電率が $\sigma=10^{-3}$ S/m 程度の場合における信号の減衰は，半径方向における半導電層の充放電現象に起因すると考えられます．$\sigma=10^3$ S/m については，半導電層の軸方向電流伝搬が減衰の主要因であると考えられます．以下では，これらを定量的に示します．

各半導電層の時定数 τ は次式で与えられます．

$$\tau = CR = \frac{2\pi\varepsilon_0\varepsilon_r}{\ln(r_2/r_1)} \frac{\ln(r_2/r_1)}{2\pi\sigma} = \frac{\varepsilon_0\varepsilon_r}{\sigma}, \quad (1)$$

ここで，$C=2\pi\varepsilon_0\varepsilon_r/\ln(r_2/r_1)$ は半導電層の単位長あたりのキャパシタンス，$R=\ln(r_2/r_1)/2\pi\sigma$ は半導電層の半径方向の単位長あたりの抵抗，r_2 は半導電層の外側半径，r_1 は半導電層の内側半径，ε_0 は真空の誘電率，ε_r は半導電層の比誘電率，σ は半導電層の導電率です．$\varepsilon_r=3$, $\sigma=10^{-3}$ S/m の場合，時定数は $\tau=27$ ns となります．導電率 $\sigma=10^{-3}$ S/m の半導電層の半径方向における充放電は，半周期17 ns の30 MHz 信号に強い影響を及ぼし，著しい減衰と変わいを引き起こします．なお，

この条件では，半導電層を半径方向に流れる伝導電流の大きさは半径方向の変位電流の大きさと同程度になります．このことは，半導電層のコンダクタンスとサセプタンスが同程度になるといい換えることもできます．

$\sigma=10^3$ S/m という高い導電率では，周波数 f の電磁波の導電率 σ，透磁率 μ_0 の物質への透過深さ d が損失計算に関連しています．透過深さは次式で与えられます．

$$d=1/\sqrt{2\pi f \sigma \mu_0}. \tag{2}$$

$\mu_0=4\pi\times10^{-7}$ H/m, $f=30$ MHz, $\sigma=10^3$ S/m の場合，式(2)は $d=2$ mm となります．これは，半導電層の厚さ(3 mm)に近い値です．したがって，軸方向電流のほとんどは，中心およびシース導体の表面ではなく，半導電層を流れざるをえなくなります．このことが信号の著しい減衰と変わいの原因となります．

B. 伝搬速度

ここでは，電力ケーブル上での 30 MHz 信号の伝搬速度について考察します．図 4 は，半導電層が絶縁体として振る舞う $\sigma\leq10^{-4}$ S/m の場合あるいは導体として振る舞う $\sigma\geq10^4$ S/m の場合における伝搬速度は約 170 m/μs となることを示しています．したがって，信号の伝搬速度は，絶縁層と半導電層の誘電率の値あるいは絶縁層の比誘電率 ε_r の値で理論的に決まり，$300/\sqrt{\varepsilon_r}=173$ m/μs となります．

半導電層の導電率 σ が 10^{-2} から 10^2 S/m の範囲にある場合には，伝搬速度は約 135 m/μs となっており，$\sigma\leq10^{-4}$ S/m および $\sigma\geq10^4$ S/m の場合の伝搬速度より低くなっています．$f=30$ MHz, $\sigma=1$ S/m (10^{-2} から 10^2 S/m の範囲の中央)の場合，式(2)は $d=65$ mm となります．この透過深さは半導電層の厚さ(3 mm)より大きいため，軸方向電流は中心導体の表面およびシース導体の内側表面を主として流れ，磁界エネルギーは中心導体の表面とシース導体の内側表面の間に蓄えられます．したがって，対応するインダクタンスは $L'=\mu_0\ln(25\text{ mm}/5\text{ mm})/2\pi$ で与えられます．式(1)より，導電率 $\sigma=1$ S/m の半導電層の時定数は $\tau=0.027$ ns であるため，中心導体表面から内側半導電層の外側表面に，シース導体の内側表面から外側半導電層の内側表面に，電荷は半径方向に即座に移動します．したがって，電界エネルギーは絶縁層に蓄えられます．これに対応するキャパシタンスは，$C'=2\pi\varepsilon_0\varepsilon_r/\ln[(25\text{ mm}-3\text{ mm})/(5\text{ mm}+3\text{ mm})]$ で与えられます．L' と C' から，対応する信号

の伝搬速度は理論的に次式のように与えられます．

$$v' = \frac{1}{\sqrt{L'C'}} = \sqrt{\frac{2\pi}{\mu_0 \ln(25/5)} \frac{\ln(22/8)}{2\pi\varepsilon_0\varepsilon_r}} = 132 \text{ m}/\mu\text{s}, \tag{3}$$

この値は，FDTD 計算で得た伝搬速度 (135 m/μs) に良好に一致します．

1.1.10　まとめ　*Summary or Conclusions*

まとめでは，実験や計算の結果から導かれた結論を簡潔に述べます．本文中で言及していないことを，まとめで述べることがないようにします．また，通常は，参考文献を引用しませんし，図や式も使用しません．

下記は，「電力ケーブル上での電力線搬送通信信号の伝搬に与える半導電層の影響」という題目の論文におけるまとめの例です．

We have studied the propagation characteristics of a PLC signal of frequency $f = 30$ MHz along a single-core power cable having two, 3-mm thick semiconducting layers using the FDTD method. We found that a PLC signal suffers significant attenuation and dispersion particularly when the semiconducting-layer conductivity σ is around 10^{-3} and 10^3 S/m. The reason for the significant signal attenuation around $\sigma = 10^{-3}$ S/m is that charging and discharging processes in a radial direction of the semiconducting layers become marked since the time constant $\tau = CR$ (=27 ns) of the semiconducting layer is close to the half cycle (17 ns) of a 30-MHz signal. In this condition, the conductance of the semiconducting layer is close to its susceptance. The reason for the significant signal attenuation around $\sigma = 10^3$ S/m is that axial current flows in the semiconducting layers because a penetration depth (2 mm) for $f = 30$ MHz and $\sigma = 10^3$ S/m is close to the thickness of the semiconducting layer (3 mm). Therefore, it is quite difficult to conduct PLC signals in a power system cable with semiconducting layers of conductivity about $\sigma = 10^{-3}$ or 10^3 S/m, but there are more possibilities if $\sigma \leq 10^{-5}$ S/m or $\sigma = 1$ S/m.

私たちは，FDTD 法を用いて，厚さ 3 mm の半導電層を 2 層有する単芯電力ケーブル上での周波数 $f = 30$ MHz の PLC 信号の伝搬特性に

ついて検討を行いました．このFDTD計算により，半導電層の導電率が10^{-3}および10^3 S/m程度のときには，PLC信号は著しく減衰することを発見しました．導電率が10^{-3} S/mの場合に生じる信号の著しい減衰は，半導電層の時定数$\tau=CR(=27 \text{ ns})$が30 MHz信号の半サイクル(17 ns)に近くなることによる半導電層の半径方向の充放電現象が原因です．このとき，半導電層のコンダクタンスはサセプタンスに近い値となっています．導電率が10^3 S/mの場合に生じる信号の著しい減衰は，周波数$f=30$ MHz，導電率$\sigma=10^3$ S/mに対する透過深さ(2 mm)が半導電層の厚さ(3 mm)に近くなることにより半導電層に軸方向電流が流れることが原因です．したがって，$\sigma=10^{-3}$または10^3 S/m程度の半導電層を有する電力ケーブルでは，PLC信号の伝導は困難となります．しかし，$\sigma \leq 10^{-5}$ S/mまたは$\sigma=1$ S/mの場合には，PLC信号伝導の可能性は高まります．

1.1.11　付録　*Appendices*

たとえば，論文中で使用した近似理論式の導出過程や使用した数値計算プログラムの計算精度や数値安定性の試験結果など，読者が論文の本文を読み進めるためには不要ですが，論文には収録しておきたい内容を付録とします．付録は，論文に必ず必要な項目ではありません．

1.1.12　謝辞　*Acknowledgments*

研究の本質的な部分に携わった方々の名は，その方々の同意を得たうえで，著者に含めます．ちょっとしたお手伝いをしてくれた方，助言をくれた方，議論してくれた方については，著者には含めず，謝辞の部分で謝意を述べます．その場合も，投稿前に論文原稿をその方々に送り，同意を得ておきましょう．査読者が，論文内容の改善に非常に役立つコメントをくれた場合には，下記のように，査読者(anonymous reviewers)への謝辞も述べることができます．

We would like to thank John Smith, and three anonymous reviewers for their valuable comments on the paper.

この論文について貴重なコメントをくださったジョン・スミスと3人

第1章　英語論文執筆法

> の匿名の査読者にお礼を申し上げます．

　所属する研究室以外の組織の特別な設備を利用した場合には，下記のように，その組織への謝辞を述べます．

> The computations in this research were performed on a supercomputer, Kyoto SUP III, of the Kyoto Computer Center, Kyoto, Japan. We would like to acknowledge the Kyoto Computer Center for permitting us to use it.
>
> この論文の計算は，京都コンピュータセンターのスーパーコンピュータ Kyoto SUP III を用いて行われました．使用を許可してくださった京都コンピュータセンターに感謝いたします．

　また，科学技術研究費などの補助を受けて実施した研究の場合には，その旨を述べます．このような場合の例を以下に示します．

> This research was supported by the Ministry of Education, Culture, Sports, Science and Technology (MEXT) of Japan under Grant 88888888.
>
> 本研究は，文部科学省の研究補助金(課題番号 88888888)を受けて行われました．

1.1.13　参考文献　*References*

　先行研究で明らかにされた知見を話題にした場合，その文献を必ず引用しましょう．ただし，委託研究報告書のような機密資料の引用は避けます．また，私信(private communication)の引用もできるだけ避けます．掲載が決定し印刷中の論文を引用する場合には，その論文情報の末尾に"(in press)"をつけておきます．論文中で引用した文献の著者名，題目，雑誌名(あるいは著書名，会議名)，巻号頁番号，発行年をすべてリストにして記載します．著者名が多い場合でも，参考文献リストにおいては，"and others"を意味する"et al."は用いずに，全著者名を記しましょう．ただし，著者数が一定以上(例えば 7 名以上)であれば，第一著者名に"et al."を付けて記すことを指

定している雑誌もあります．参考文献は，適切な査読者を選定しなければならない編集者にとっても，関連する研究に興味をもった読者にとっても，重要な情報となります．したがって，ミスタイプしないように注意深く記載する必要があります．また，ほかの研究者の成果や業績を認め評価していることを示す(to give credit)意味もありますので，著者自身の過去の論文の引用は必要最小限にとどめ，関連する研究者の文献を偏りなく引用しましょう．

参考文献の書き方には，バンクーバー方式(Vancouver referencing system)とハーバード方式(Harvard referencing system)があります．

バンクーバー方式では，下記のように，論文本文での引用順に番号をつけ参考文献を列挙します．本文の引用箇所の直後には，[1]や(1)あるいはそれらの上付き文字を挿入します．

[1] Y. Baba, N. Tanabe, N. Nagaoka, and A. Ametani, "Transient analysis of a cable with low-conducting layers by the finite-difference time-domain method," IEEE Transactions on Electromagnetic Compatibility, vol. 46, no. 3, pp. 488–493, Aug. 2004.

[2] K. Steinbrich, "Influence of semiconducting layers on the attenuation behaviour of single-core power cables," IEE Proceedings on Generation, Transmission & Distribution, vol. 152, no. 2, pp. 271–276, Mar. 2005.

[3] A. Cataliotti, A. Daidone, and G. Tine, "Power line communication in medium voltage systems: characterization of MV cables," IEEE Transactions on Power Delivery, vol. 23, no. 4, pp. 1896–1902, Oct.

表 1.2 学術論文中で用いられるラテン語の略語ともとのラテン語，対応する英語

略　語	ラテン語	英　語
cf.	confer	compare
C.V.	curriculum vitae	résumé
e.g.	exempli gratia	for example
et al.	et alia	and others
etc.	et cetera	and so on
ibid.	ibidem	in the same place
i.e.	id est	that is
vs.	versus	against

> 2008.

　ハーバード方式では，下記のように，論文本文で引用した文献の第1著者の姓のアルファベット順(alphabetical order)に列挙します．このとき，第1著者のファーストネームとミドルネームのイニシャル文字は姓の後ろに記します．本文の引用箇所の直後には，たとえば，"(Baba et al. 2004)"や"[Steinbrich 2005]"のようにして，その文献の著者名と発行年を括弧で囲んだものを挿入します．通常，引用した文献の著者が2名までは，引用箇所でも全員の姓を記載し，3名以上の場合は，第1著者名と"et al."を用いて記載します．同一の著者による複数の文献を引用する場合には，古い発行年のものから並べます(chronological order)．同一の著者による同一年に発行された異なる複数の論文を引用する場合には，(2004a)，(2004b)，...として区別します．

> Baba, Y., N. Tanabe, N. Nagaoka, and A. Ametani (2004), "Transient analysis of a cable with low-conducting layers by the finite-difference time-domain method," IEEE Transactions on Electromagnetic Compatibility, vol. 46, no. 3, pp. 488–493.
> Cataliotti, A., A. Daidone, and G. Tine (2008), "Power line communication in medium voltage systems: characterization of MV cables," IEEE Transactions on Power Delivery, vol. 23, no. 4, pp. 1896–1902.
> Steinbrich, K. (2005), "Influence of semiconducting layers on the attenuation behaviour of single-core power cables," IEE Proceedings on Generation, Transmission & Distribution, vol. 152, no. 2, pp. 271–276.

　"et al."はラテン語"et alia"の略語です．学術論文でよく用いられるラテン語の略語ともとのラテン語，対応する英語を**表1.2**に示します．ほとんどの場合，これらの略語は論文中で定義することなしで使用できます．

　Microsoft社のWordを用いて論文を執筆する場合には，「参考資料」タブの「引用文献と文献目録」グループに文献引用や参考文献リスト作成に役立つ機能があります．バンクーバー方式のIEEEスタイルやハーバード方式などの書式も選択することができます．

1.1.14 著者略歴　*Author Biographies*

論文中の著者の略歴欄には，学歴，職歴，関連分野の受賞歴，関連学会の委員歴，関連学会の会員情報などを記します．

> Hanako Denki (Student Member) received the B.Sc. and M.Sc. degrees, both in Electrical Engineering, from the University of Southern Kyoto, Kyoto, Japan, in 2007 and 2009, respectively. She is currently a Ph.D. student of the same University, and also a research fellow of the Japan Society for the Promotion of Science. She is the recipient of the Outstanding Master's Thesis Award from the University of Southern Kyoto. She is a Student Member of the Americal Geophysical Union.
>
> 電気花子(学生員)は，2007年および2009年に日本の南京都大学から電気工学の学士号および修士号を取得しました．現在，彼女は同大学の博士課程の学生で，日本学術振興会の特別研究員を兼任しています．彼女は南京都大学の優秀修士論文賞の受賞者です．彼女はアメリカ地球物理学連合の学生員です．

理学のみではなく，理系の学士号および修士号は "Bachelor of Science degree" および "Master of Science degree" といい，それぞれ "B.Sc. degree" または "B.S. degree"，"M.Sc. degree" または "M.S. degree" と略記されます．文系理系に関係なく，博士号は "Ph.D." と記されます．これは，ラテン語の "Philosophiae Doctor"（英語の "Doctor of Philosophy"）の略記です．学位名については，付3.7の補足で詳しく説明しています．

1.2　能動態と受動態の使い方
How to Use Active and Passive Voices

本節では，学術論文中で用いる能動態と受動態について説明します．昔は，著者自身を主語とした能動態の文は避けられていたようですが，最近では，著者自身を主語(we, the authors など)にした能動態の文も積極的に用いられています．

1.2.1 能動態と受動態の関係
Relation Between Active and Passive Voices

他動詞を含む能動態の文のほとんどは，形式的には受動態に変換することができます．たとえば，下記の "We" を主語とし，"the voltage drop across each resistor" を目的語とする能動態の文は，"The voltage drop..." を主語にした受動態の文に書き換えることができます．

> We measured the voltage drop across each resistor.
>
> 私たちは各抵抗での電圧降下を測定した．
>
> The voltage drop across each resistor was measured by us.
>
> 各抵抗での電圧降下が私たちによって測定された．

これらの二つの文は同じ内容を表していますが，"We" を主語とする文では，測定を行った動作主(私たち)に，"The voltage drop across each resistor" を主語とする文では，測定という動作の対象(各抵抗での電圧降下)に重点が置かれている点が異なっています．なお，他動詞のなかには，受動態に変換できないものもあります(have, possess, last, lack, fit, resemble など)．

また，形式的に受動態に変換すると，もとの能動態の文と意味が変ってしまう場合もあります．たとえば，下記の能動態の文では，"two languages" は学生によって異なる可能性がありますが(ある学生は日本語と英語を，別の学生は中国語と英語を話すかもしれない)，受動態の文では，"two languages" は特定の2カ国語という意味になります．

> Every student in our laboratory speaks two languages.
>
> 私たちの研究室では，どの学生も2カ国語を話します．
>
> Two languages are spoken by every student in our laboratory.
>
> 私たちの研究室では，2カ国語がどの学生によっても話されています．

以上の2例からも明らかなように，変換した受動態の文がもとの能動態の

1.2.2 能動態の使い方　*How to Use Active Voice*

下記の例からも明らかなように，受動態を用いた場合には，主語の部分が長い頭でっかちの文になってしまうのに対して，能動態を用いた場合には，主語の部分が短くなり明瞭な文になるという利点があります．このため，最近の学術論文では，著者自身を主語にした能動態の文も積極的に用いられています．

> The propagation characteristics of a 30-MHz PLC signal along a single-core power cable having two, 3-mm thick semiconducting layers were computed (by us) using the finite-difference time-domain (FDTD) method.
>
> 時間領域有限差分(FDTD)法を用いて，厚み3 mmの半導電層を2層有する単芯電力ケーブル上での周波数30 MHzのPLC信号の伝搬特性の計算が(私たちにより)行われました．
>
> We computed the propagation characteristics of a 30-MHz PLC signal along a single-core power cable having two, 3-mm thick semiconducting layers using the finite-difference time-domain (FDTD) method.
>
> 私たちは，時間領域有限差分(FDTD)法を用いて，厚み3 mmの半導電層を2層有する単芯電力ケーブル上での周波数30 MHzのPLC信号の伝搬特性の計算を行いました．

著者自身を表す主語としては，著者が複数の場合には，"we"や"the authors"が一般的に用いられています．ただし，同一の論文のなかで，"we"と"the authors"が混用されることはありません．単著論文の場合には，これまでのところ，電気電子工学分野の雑誌においては"I"はほとんど用いられておらず，"we"または"the author"が用いられています．単著論文で"we"を使用するのは奇妙ですので，何年か後には，"I"が堂々と用いられているかもしれません．

下記の第2文のように，無生物主語の能動態の文も，受動態の文よりも明瞭であるため，よく用いられています．

第 1 章　英語論文執筆法

> FDTD-computed waveforms of the voltage between the core and sheath conductors of the cable at distances of 20, 40, 60, 80, and 100 meters from the excitation point are shown in Figure 1.
>
> 電圧印加点から 20，40，60，80 および 100 m 離れた点におけるケーブル中心導体とシース導体間電圧の FDTD 計算波形が図 1 に示されています．
>
> Figure 1 shows FDTD-computed waveforms of the voltage between the core and sheath conductors of the cable at distances of 20, 40, 60, 80, and 100 meters from the excitation point.
>
> 図 1 は，電圧印加点から 20，40，60，80 および 100 m 離れた点におけるケーブル中心導体とシース導体間電圧の FDTD 計算波形を示しています．

1.3　時制の選び方
How to Choose Tense

　本節では，学術論文中で用いる時制について説明します．時制には，現在，過去，未来の三つの基本時制があり，それらのそれぞれに完了形があります．さらに，それらのすべてに進行形がありますので，合計 12 の時制が存在します．学術論文中においては，読者がその論文を読んでいる瞬間を現在と考え，それを基準にして時制を選択します．したがって，現在時制を用いる箇所が最も多くなります．現在完了時制，過去時制，未来時制も用いられます．

1.3.1　現在時制の使い方　*How to Use Present Tense*

　学術論文中においては，読者がその論文を読んでいる瞬間を現在と考え，それが基準の時制となります．したがって，現在時制を用いる箇所が最も多くなります．

　緒言(Introduction)での論文構成の説明，研究方法(Method)での計算モデルの説明，および研究結果(Results)での実験結果や計算結果の説明の例

文を以下に示します.

> This paper is organized as follows（または The structure of this paper is as follows）. In Section 2, we present a method for computing electric and magnetic fields in the vicinity of a lightning strike object using the FDTD method. In Section 3, we show the computed results for currents along the lightning strike object and the close electric and magnetic fields for different values of ground conductivity…
>
> この論文は，次のように構成されています．第2節では，雷撃構造物近傍の電界および磁界をFDTD法により計算するための方法を示します．第3節では，異なるいくつかの大地導電率に対する雷撃高構造物に沿った電流およびその構造物近傍の電界，磁界の計算結果を示します…
>
> Figure 1 shows a single-core power cable, to be analyzed using the FDTD method. The radius of the core conductor is 5 mm. The inner radius of the sheath conductor is 25 mm. Both the core and sheath conductors are perfectly conducting.
>
> 図1は，FDTD法を用いた解析に使用する単芯の電力ケーブルを示しています．中心導体の半径は5 mmです．シース導体の内半径は25 mmです．中心導体およびシース導体はともに完全導体です．
>
> Figure 4 shows the dependence of the magnitude of the voltage between the core and sheath conductors on the semiconducting-layer conductivity at distance 100 m from the excitation point. It follows from Figure 4 that the dependence of the signal attenuation on the semiconducting-layer conductivity is not monotonic.
>
> 図4は，電圧印加点から100 mの距離における中心導体-シース導体間電圧の大きさの半導電層導電率への依存性を示しています．図4より，信号の減衰の半導電層の導電率への依存性は単調ではないことがわかります．

なお，現在の状態や事実だけではなく，実現が確実な未来のことも現在時

制で表すことができるため，上記の緒言における例文の第 2 および第 3 文は，次節および次々節に関する未来のことを現在時制で記述しています．もちろん，これらの 2 文は，"In Section 2, we will present ... In Section 3, we will show ..." と未来時制で書くこともできます．

語や語群の省略時に使う "..." は "ellipsis" といいます．

1.3.2　未来時制の使い方　*How to Use Future Tense*

読者が読んでいる部分より先の部分や今後の研究課題などについて述べる場合に，未来形を用います．未来形の使用例を以下に示します．

> In the following section, we will use the FDTD method to simulate a lightning strike to a tall structure.
>
> 次節においては，高構造物への雷撃を模擬するために，FDTD 法を使用します．

なお，実現が確実な未来のことは現在時制で表すこともできるため，この例文を，"In the following section, we use the FDTD method ..." と現在時制で書くこともできます．

1.3.3　過去時制の使い方　*How to Use Past Tense*

過去の論文の結果や，その論文において読者がすでに読み終えている部分について述べる場合には過去時制を用います．過去時制の使用例を以下に示します．

> In 1967, Newman et al. (1967) performed an experiment for triggering a lightning discharge from a natural thundercloud.
>
> 1967 年に，ニューマンらは，自然の雷雲から雷放電を誘発する実験を行いました．
>
> In the preceding section, we discussed the mechanism of voltage attenuation.

> 前節において，私たちは電圧減衰のメカニズムについて検討を行いました．

1.3.4　現在完了形の使い方　*How to Use Present Perfect Tense*

現在完了時制は，過去の実験，計算あるいは発見などが，現在にも影響していることを表す場合に用いられます．例を以下に示します．

> Newman et al. have performed an experiment for triggering a lightning discharge from a natural thundercloud.
>
> ニューマンらは，自然の雷雲から雷放電を誘発する実験を行いました．

ただし，時期を明記する場合には，"In 1967, Newman et al. performed an experiment for triggering a lightning discharge from a natural thundercloud."と過去時制を用いなければなりません．過去時制で表現すると，その実験が現在にも影響しているという感じがなくなってしまいます．論文中で先行研究について紹介する場合，初回は現在完了形を用い，2回目以降それについて言及する際は過去形を用います．

まとめ(Summary)や結論(Conclusions)の部分で，下記のように，現在完了時制の完了・結果の用法を用いて，今まさに本研究を終えたという感じを出すことができます．

> In this paper, we have studied the propagation characteristics of a 30-MHz PLC signal along a single-core power cable using the FDTD method.
>
> 本論文において，私たちは，FDTD法を用いて，単芯電力ケーブル上における30 MHzのPLC信号の伝搬特性について研究を行いました．

1.3.5　時制の一致の例外
　　　Exceptions to the Rule of Sequence of Tenses

下記のように，主節の動詞または助動詞の時制を過去にすると，従属節の

動詞または助動詞の時制も現在から過去あるいは過去から過去完了に変化させる必要があります．このことを時制の一致といいます．

> We think that this method is useful.
>
> 私たちは，この方法が有用であると思っています．
>
> We thought that this method was useful.
>
> 私たちは，この方法が有用であると思っていました．

しかし，下記のように，従属節が不変の真理や現在の事実を示す場合には，時制の一致は適用されず，従属節においては現在時制が用いられます．

> Einstein discovered that energy and mass are equivalent.
>
> アインシュタインは，エネルギーと質量が等価であることを発見しました．
>
> We found that the voltage attenuation is significant at frequency 3 GHz.
>
> 私たちは，周波数 3 GHz において電圧の減衰が著しいことを発見しました．

なお，これらの従属節の "are" を "were" に，"is" を "was" にしても間違いではありませんが，そのようにすると，「エネルギーと質量が等価である」，「3 GHz において電圧の減衰が著しい」ことが現在においても正しい不変の事実であるという感じがなくなってしまいます．

1.4　助動詞の使い方
How to Use Auxiliary Verbs

本節では，学術論文中での助動詞の使い方について説明します．"would"，"could"，"might"，"should" は，それぞれ "will"，"can"，"may"，"shall" の

過去形ですが，現在の意味を表す場合にもよく用いられます．学術論文において，単純未来や主語の意志や推量を表す場合には，"will" が用いられるのがほとんどで，"shall" や "be going to" が用いられることはほとんどありません．"be able to" は人間が主語の文でのみ用いられます．また，その過去形は実際にそのようなことができたという事実を意味するのに対し，"could" はそのような能力があったことを意味し，両者は完全に等価というわけではありません．

1.4.1 "will" と "would" の使い方
How to Use "will" and "would"

"will" は，単純未来，現在の習慣，主語の意志や推量を表す場合に用いられる助動詞です．イギリス英語では，単純未来や主語の意志を表すのに "shall" が用いられることがありますが，学術論文中では，この目的で "shall" が用いられることはほとんどありません．"be going to" も学術論文中では，ほとんど用いられることはありません．"will" の使用例を以下に示します．

> In the following section, we will use the FDTD method to simulate a lightning strike to a tall structure.
>
> 次節においては，高構造物への雷撃を模擬するために，FDTD 法を使用します．

"would" は，"will" の過去形で，過去における主語の意志や過去の不規則な習慣を表す場合に用いられます．また，下記のように，丁寧な表現，慣用表現 (would like)，仮定法の含みのある表現で用いられる "would" は過去形ですが，意味は現在です．

> Would you please turn off the lights in this room?
>
> すみませんが，この部屋の照明を消していただけませんでしょうか．
>
> I would like to talk about our system for measuring lightning currents.
>
> 私たちの雷電流測定システムについてお話したいと思います．

> A careful reviewer would notice the error.
>
> 注意深い査読者なら，その間違いに気づくことでしょう．

下記のように，事実を単純に推定する場合にも用いられます．

> This system would allow us to measure nanoampere currents.
>
> このシステムにより，ナノアンペアの電流を測定することが可能になります．

1.4.2 "can" と "could" の使い方
How to Use "can" and "could"

"can" は，能力，推量，許可を表す場合に用いられる助動詞です．"be able to" や "be capable of...ing" も能力を表しますが，"be able to" は人間が主語の場合においてのみ使用できます．"can" の使用例を以下に示します．

> This thermometer can measure body temperature without physical contact.
>
> この温度計は体温を非接触で測定することができます．

"could" は，"can" の過去形です．下記のように，"could" と "was or were able to" は等価ではなく，"could" は「そのような能力があった」ことを意味し，"was or were able to" は「実際にそのようなことができた」という事実を意味します．

> He could write good papers in English.
>
> 彼は英語でよい論文を書く能力があった．
>
> He was able to write his Master's thesis in English.
>
> 彼は英語で修士論文を書くことができた．

推量を表す場合の例を以下に示します．

> It cannot be true.
>
> それは真実のはずがない．

"cannot" は強い否定を表します．学術論文においては，"can't" など否定の短縮形は用いられません．「それは真実に違いない．」は "It can be true." ではなく "It must be true." といいます．過去のことに対する推量を表す場合の例を以下に示します．

> He cannot have completed it by himself.
>
> 彼が1人でそれを完成させたはずがない．

"cannot" を "could not" に変えても，ほとんど同じ意味になります．

許可を表す "can" は口語表現ですので，学術論文ではほとんど用いられませんが，例を以下に示します．

> You can use my personal computer.
>
> 私のパソコンを使ってもいいですよ．

下記のように，丁寧な表現，仮定法の含みのある表現で用いられる "could" は過去形ですが，意味は現在です．

> Could you please turn off the lights in this room?
>
> すみませんが，この部屋の照明をすべて消していただけませんでしょうか．
>
> Perhaps she could solve this difficult problem.
>
> 彼女なら，この難問を解くことができるだろう．

1.4.3 "may" と "might" の使い方
How to Use "may" and "might"

"may" は，下記のように，許可，推量，祈願を表す場合に用いられる助動詞です．

> May I use this computer?
>
> このコンピュータを使ってもよろしいですか．
>
> He may not be aware of this error.
>
> 彼はこの間違いに気づいていないかもしれない．
>
> May you be very happy!
>
> ご多幸をお祈りします．

"might" は，"may" の過去形です．時制の一致で "may" の過去形として用いられる以外では，下記のように丁寧な表現および仮定法の含みのある表現でのみ用いられます．

> Might I ask you questions?
>
> 質問をしてもよろしいでしょうか．
>
> We might buy the computer if it cost about 3 000 dollars.
>
> 3 000 ドルぐらいでなら，そのコンピュータを買うかもしれない（実際には買わない）．
>
> I might have died in the traffic accident.
>
> その交通事故で死んでいたかもしれない（実際には死ななかった）．

〈補足〉 上記の第 2 文は仮定法過去，第 3 文は仮定法過去完了の構文です．

1.4.4 "must" の使い方　*How to Use "must"*

"must" は(話し手が課す)必要, 推量を表す場合に用いられる助動詞です. 外部的事情による必要性を表す場合には, "have to" を用います. 必要を表す場合の使用例を以下に示します.

> I must finish this paper work by tomorrow.
>
> 私は, 明日までに, この事務処理を終えなければなりません.
>
> I have to pay the publication fee for this paper by next Tuesday.
>
> 私は, 今度の火曜日までに, この論文の掲載料を支払わなければなりません.

過去の必要を表す場合には, "must" には過去形がないため, "had to" を代用します.

推量を表す場合の例を以下に示します.

> It must be true.
>
> それは真実に違いない.

「それは真実のはずがない.」は, "It must not be true." ではなく, "It cannot be true." といいます. "must not" は強い禁止を表します.

過去のことに対する推量を表す場合の例を以下に示します.

> It must have been easy for her to earn her Ph.D. degree.
>
> 彼女にとっては, 博士号を取得するのは容易であったに違いない.

1.4.5 "shall" と "should" の使い方
How to Use "shall" and "should"

イギリス英語では, 単純未来や主語の意志を表すのに "shall" が用いられ

ることがありますが，学術論文中では，この目的で "shall" が用いられることはほとんどありません．一方，規則，法律あるいは規格文書においては，"shall" が "must"（強い義務）の意味で用いられます．

> All passengers shall wear seat belts.
>
> すべての乗客はシートベルトを着用すること．
>
> The indirect lightning performance calculation shall take into account the effect of the ground conductivity.
>
> 間接雷に対する性能を計算する場合には，大地導電率を考慮に入れること．

"should" は，"shall" の過去形ですが，下記のように，現在の義務を表すのによく用いられます．

> You should follow his advice.
>
> あなたは，彼のアドバイスに従うべきです．
>
> You should not jump to conclusions.
>
> 急いで結論を出してはいけません．

"should" の義務を表す意味の強さは，"shall"，"must"，"had better" ほど強くはありません．

過去において実行されなかった行為について述べる場合の例を以下に示します．

> I should have listened to you.
>
> あなたのいっていることを聞いておくべきでした．

1.5 不定冠詞と定冠詞の使い方
How to Use Indefinite and Definite Articles

　本節では，学術論文中での冠詞の使い方について説明します．不定冠詞（a または an）は，単数形の可算名詞の初出時に，それが読者にとって不特定のものである場合，その名詞の前につけられます．読者にとって不特定の複数形の可算名詞には，冠詞はつけられません．定冠詞（the）は，初出，既出，可算，不可算，単数形，複数形に関係なく，名詞がなにを指しているかが著者だけではなく読者にとっても明らかな場合に，その名詞の前につけられます．「図1」は，固有名詞的に用いられ，冠詞をつけずに最初の文字を大文字にして，"Figure 1（または短縮形で Fig. 1）"と書かれます．

1.5.1 不定冠詞の使い方 *How to Use Indefinite Articles*

　不定冠詞（a または an）は，単数形の可算名詞の初出時に，それが読者にとって不特定なものである場合，その名詞の前につけられます．たとえば，下記のように，提案する "model" は，著者にとっては特定のものですが，読者にとっては初出時にはどのようなものかわからない不特定のものですので，不定冠詞がつけられています．

> In this paper, we propose a model of corona discharge for surge computations using the finite-difference time-domain (FDTD) method.
>
> この論文では，私たちは時間領域有限差分（FDTD）法を用いたサージ計算のためのコロナ放電モデルを提案しています．

　同格の "of" を用いて，物理量を具体的な数値で表す場合には，下記のように物理量に不定冠詞をつけます．具体的な数値を伴うと定まった感じになり，定冠詞をつけてしまいそうになりますが，いくつかあるなかの一つという意味に理解しましょう．

> a current of 10 A（amperes）

> 10 A の電流
>
> an impedance of 70 Ω (ohms)
>
> 70 Ω のインピーダンス
>
> a temperature of 300 K (kelvins)
>
> 300 K の温度
>
> a distance of 2 km (kilometers)
>
> 2 km の距離

"a"か"an"の選択は，下記のように，冠詞の直後にくる語が子音の発音で始まるか，母音の発音で始まるかで決まります．可算名詞である"computation"を修飾する形容詞"finite"の発音は「ファイナイト」(子音)ですので，この場合には，"a"が用いられています．一方，頭字語"FDTD"の発音は「エフディティディ」(母音)ですので，"an"が用いられています．

> a university
>
> 大学
>
> a finite-difference time-domain computation
>
> 時間領域差分計算
>
> an FDTD computation
>
> 時間領域差分計算

1.5.2 定冠詞の使い方　*How to Use Definite Articles*

定冠詞(the)は，可算，不可算，単数形，複数形に関係なく，既出の名詞

を指す場合に，その名詞の前につけられます．たとえば，下記の第2文の"model"は，第1文の"a model"を指しているため，定冠詞がつけられています．

> In this paper, we propose a model of corona discharge for surge computations using the finite-difference time-domain (FDTD) method. We compare the waveforms of corona current, computed with the model, with the corresponding measured waveforms.
>
> この論文では，私たちは時間領域有限差分(FDTD)法を用いたサージ計算のためのコロナ放電モデルを提案しています．そのモデルを用いて計算したコロナ電流の波形を，対応する実測波形と比較しています．

初出であっても，それがなにを指しているかが読者にとって明らかな場合には，定冠詞がつけられます．たとえば，上の例文において，"finite-difference time-domain (FDTD) method"は初出ですが，この論文を投稿した雑誌の読者のほとんどが知っているような方法である場合には，このように定冠詞がつけられます．

形容詞句や形容詞節により名詞が限定され，さらに特定されている場合にも，定冠詞がつけられます．たとえば，上の例文の第2文の"waveforms"は初出ですが，「(提案した)モデルを用いて計算したコロナ電流の」という形容詞句により特定されており，読者にとってもそのことが明らかですので，定冠詞がつけられています．"corresponding measured waveform"も初出ですが，「(特定の)計算波形に対応する実測波形」という意味であり，なにを指しているかが読者にとっても明らかですので，定冠詞がつけられています．ただし，下記のように，「カメラ内蔵の」という形容詞句で限定されていても，数あるカメラ内蔵のノートパソコンのうちの不特定の一つを指す場合には，不定冠詞がつけられます．

> a notebook-size personal computer with a built-in camera
>
> カメラ内蔵の(不特定のある一つの)ノートパソコン

序数詞，最上級の形容詞や"only"，"same"によって特定されている場合にも，定冠詞がつけられます．

なお，冠詞の直後にくる語が子音の発音で始まる場合には，"the"を「ザ」

と発音し，母音で始まる場合には，「ズィ」と発音します．たとえば，"the finite..."は，「ザ　ファイナイト...」と発音しますが，"the FDTD method"は「ズィ　エフディティディ　メソッド」と発音します．

1.5.3　無冠詞の使い方　*How to Use Nouns Without Articles*

複数形の可算名詞が特定のものを表していない場合，その名詞には定冠詞はつけられません．たとえば，1.5.2の例文において，"surge computations"は直後の形容詞句"using the finite-difference time-domain (FDTD) method"により，「時間領域有限差分(FDTD)法を用いたサージ計算」と限定されてはいますが，時間領域有限差分法を用いた数あるサージ計算のなかの不特定のいくつかを意味しているだけで，特定されてはいませんので，無冠詞の複数形として用いられています．

同じ例文の"a model of"の直後の"corona discharge"および"the waveforms of"の直後の"corona current"が無冠詞である理由は，「コロナ放電」および「コロナ電流」という現象を抽象的に示しているためです．

表1，図1，式(1)は，固有名詞的に用いられ，冠詞をつけずに最初の文字を大文字にして，"Table 1"，"Figure 1(または短縮形でFig. 1)"，"Equation (1)〔またはEq. (1)〕"と書かれます．なお，"Fig. 1"や"Eq. (1)"では，ピリオドは終止符としてではなく，短縮や省略を表す記号として用いられています．このピリオドの用法は，例えば，"Doctor"が"Dr."，"Number"が"No."，"Hanako Denki"がイニシャルで"H. D."と表されているように，広く用いられています．短縮や省略を表すピリオドが文末にくる場合，"Her initials are H.D.."とは表記せず，終止符の役割も兼ねさせて，"Her initials are H.D."と表記します．

"Euler's formula"（オイラーの式），"Ampere's law"（アンペアの法則），"Planck's constant"（プランク定数），"Maxwell's equations"（マクスウェル方程式）のように，人名の語尾にアポストロフィエスをつけ所有格で用いる場合には，固有名詞扱いとなり冠詞をつけません．アポストロフィエスをつけずに定冠詞をつけて，"the Maxwell equations"という場合もありますが，"Maxwell's equations"がより頻繁に用いられています．この場合，"Maxwell"という固有名詞により"equations"が特定されるため，"the"がつくと考えます．"the Maxwell's equations"とはいいません．

1.6 接続詞，分詞構文，関係代名詞の使い方
How to Use Conjunctions, Participial Constructions, and Relative Pronouns

　本節では，学術論文中での接続詞，分詞構文および関係代名詞の使い方について説明します．接続詞は，語，句あるいは節を連結する語です．分詞構文は，条件，譲歩，理由などを表す副詞節における接続詞と主語を省き，それらを動詞の分詞に置き換えた副詞句を含む文です．文は簡潔になりますが，接続詞と主語が省略されるため，学術論文において使用すると，明確さが失われる場合もあります．関係代名詞は，接続詞の役割を兼ねた代名詞のことです．関係代名詞には，限定用法と継続用法があります．

1.6.1 接続詞の使い方　*How to Use Conjunctions*

　接続詞は，語，句あるいは節を連結する語です．接続詞には，文法上対等の関係にある語，句あるいは節を連結する "and"，"or" などの等位接続詞 (coordinate conjunction) と，一方の節を他方の節に従属させる関係で連結する "that"，"since"，"although" などの従属接続詞 (subordinate conjunction) があります．

　まずは，等位接続詞 "and" および "but" の使用例を以下に示します．

> With increasing temperature, most conductors increase in resistance and insulators decrease in resistance.
>
> 　温度が上昇するにつれて，ほとんどの導体の抵抗は上昇し，絶縁体の抵抗は低下します．
>
> We tried to calculate the inductance value of this coil, but we failed to do so.
>
> 　このコイルのインピーダンスを計算しようとしましたが，失敗しました．

　ただし，"We tried to calculate the impedance of this coil. But we failed to do so." のように等位接続詞である "But" や "And" を文頭に置いた文は，文法上は誤りですので，望ましくはありませんが，最近は学術論文中でも用いられています．また，学術論文中では，等位接続詞 "but" はあまり用いられず，従属接続詞の "although" あるいは副詞の "however" を用いた下記のような文がより頻繁に用いられます．

> Although we tried to calculate the inductance value of this coil, we failed to do so.
>
> We tried to calculate the inductance of this coil. However, we failed to do so.
>
> このコイルのインダクタンスを計算しようとしましたが，失敗しました．

"both ... and 〜"（...と〜の両方とも），"neither ... nor 〜"（...も〜もない），"not only ... but (also) 〜"（...だけではなく〜も）などのように一対の語句で等位接続詞の働きをするものは等位相関接続詞と呼ばれており，学術論文中においてもよく用いられます．これらの使用例を以下に示します．

> The loss in a transmission line is caused by both the parallel conductance and the series resistance.
>
> 伝送線路の損失は並列コンダクタンスと直列抵抗の両方が原因です．
>
> A current pulse propagates along this coaxial cabe with neither attenuation nor dispersion.
>
> この同軸ケーブルを，電流パルスは減衰も分散もしないで伝搬します．
>
> This instrument can be used not only in direct-current (DC) circuits but also in alternating-current (AC) circuits.
>
> この計器は直流回路だけではなく交流回路にも使うことができます．

最後の "not only ... but (also) 〜" の文は，群前置詞 "〜as well as ..." を用

いて，"This instrument can be used in AC as well as DC circuits." のように同じ意味の文を表現することができます．

次に，従属接続詞の使用例を示します．下記は，名詞節を導く "that" および "whether(＝if)" の使用例です．

> We have proven that vector A is equal to vector B.
>
> 私たちはベクトル A がベクトル B に等しいことを証明しました．
>
> We doubt whether this method can be applicable to electromagnetic field computations.
>
> 私たちは，この方法が電磁界計算に適用できるかどうかを疑問に思っています．

下記は，理由を表す副詞節を導く "since" の使用例です．

> Since a ferromagnetic core possesses resistance, it is heated by eddy currents.
>
> 強磁性体コアは抵抗を有しているので，渦電流により熱せられます．

学術論文中では，理由を表す接続詞として "since" が最もよく用いられ，"because" もそれにつぐ頻度で用いられます．しかし，理由を表す従属接続詞 "as" はあまり用いられず，理由を表す等位接続詞 "for" もあまり用いられません．

下記は，時を表す副詞節を導く "when" および "until(＝till)" の使用例です．

> When an ammeter is required to measure a high current, a proportion of the current is diverted through a low-value resistance connected in parallel with the meter.
>
> 大電流を計測する必要があるときには，電流計に並列に接続された低抵抗に電流の一部が流されます．

> Until Faraday showed that a time varying magnetic field generates an electric field, it was thought that the electric and magnetic fields were distinct and uncoupled.
>
> ファラデーが時間変化する磁界が電界を発生することを示すまでは，電界と磁界は別々の関連のないものと考えられていました．

上記第2文の主節の動詞が過去形(showed)であるのに従属節の動詞が現在形(generates)になっているのは，従属節が現在においても正しい「不変の真理」を示しているためです．

下記は，条件を表す副詞節を導く"if"の使用例です．

> A positive contribution to the flux occurs if A has a component in the direction of dS out from the surface.
>
> A が，dS の表面から流出する方向成分をもっている場合には，その流束に対して正の寄与が生じます．

下記は，譲歩を表す副詞節を導く"although(＝though)"の使用例です．

> The most common source of lightning is the electric charge separated in ordinary thunderclouds although other types of clouds may occasionally produce natural lightning.
>
> ほかの種類の雲もときには自然雷を起こしますが，雷の最も一般的な源は通常の雷雲内で分離された電荷です．

下記は，制限を表す副詞節を導く"as far as"の使用例です．

> As far as we know, this problem has not yet been solved.
>
> 私たちの知るかぎりでは，この問題はまだ解決されるにはいたっていません．

"As far as we know"の副詞節は，"To the best of our knowledge"という副詞句に置き換えることができます．

下記は，様態を表す副詞節を導く"as if"の使用例です．

> In the simplest model, return stroke surge current distributes as if there was a vertical transmission line of constant characteristic impedance and slow propagation velocity placed above the strike object.
>
> その簡易モデルにおいては，雷撃を受けた物体の上に，特性インピーダンスが一定で，遅い伝搬速度をもつ垂直な伝送線路があるかのように，帰還雷撃サージ電流は分布します．

"as if"で導かれる副詞節は仮定法過去形となっています．
下記は，目的を表す副詞節を導く"so that"の使用例です．

> The amount of this point charge must be negligibly small so that its force will not displace nearby charges.
>
> この点電荷の電荷量は，その近くの電荷を動かすことのないぐらいに無視できるほど小さくなければなりません．

下記は，結果を表す副詞節を導く"so that"の使用例です．通常，"so that"の前にカンマが置かれます．

> This drift increases the magnitude of both the contact potential and the depletion layer thickness, so that only very few majority carriers have sufficient energy to cross the junction.
>
> このドリフトは，接触電位および空乏層の厚みの両方を大きくします．その結果，多数キャリヤのほとんどは，接合部を横切るのに十分なエネルギーをもちません．

1.6.2　分詞構文の使い方
How to Use Participial Constructions

分詞構文は，条件，譲歩，理由などを表す副詞節における接続詞と主語を省き，それらを動詞の分詞に置き換えた副詞句を含む文です．文は簡潔にな

りますが，接続詞と主語が省略されるため，学術論文において使用すると，明確さが失われる場合もあります．なお，分詞とは動詞の現在分詞形(present participle)と過去分詞形(past participle)のことで，形容詞のような働きをします．

分詞構文の使用例を以下に示します．

Using Stokes' theorem (＝if we use Stokes' theorem), we can convert the line integral of the electric field to a surface integral of the curl of the electric field.

ストークスの定理を用いると，電界の線積分を電界の回転の表面積分に変換することができます．

Trying to calculate the inductance value of this coil (＝Although we tried to calculate the inductance value of this coil), we failed to do so.

このコイルのインピーダンスを計算しようとしましたが，失敗しました．

Not having good scientific electronic calculators (＝Since I did not have any good scientific electronic calculators), I had difficulty doing the complicated calculations.

高性能な関数電卓をもっていなかったので，これらの複雑な計算を行うのに苦労しました．

Written in simple English, this textbook is read by many students.
(＝Since this textbook is written in simple English, it is read by many students.)

この教科書は平易な英語で書かれていますので，多くの学生に読まれています．

否定の分詞構文では，"not"が分詞の前に置かれます(上記第3例文)．また，受動態の分詞構文においては，多くの場合，"Being"や"Having been"は省略されます(上記第4例文)．

上記の4例では，主節の主語と従属節の主語が一致していますが，それらが一致していない場合は，従属節の主語が分詞の前に置かれます．このような構文を独立分詞構文(absolute participial construction)といいます．独立分詞構文の使用例を以下に示します．

> At frequencies greater than 1 000 MHz, transmission lines are usually in the form of a waveguide which may be regarded as coaxial lines without the center conductor, the energy being (=and the energy is) launched into the guide by probes or loops projecting into the guide.
>
> 1 000 MHz を超える周波数においては，伝送線路は，通常，中心導体のない同軸線路とみなされる導波路の形態となります．そして，導波路に突き出たプローブあるいはループにより，エネルギーが導波路に送り出されます．

この例では，主節の主語が "transmission lines"，従属節の主語が "the energy" です．

1.6.3 関係代名詞の使い方　*How to Use Relative Pronouns*

関係代名詞は，接続詞の役割を兼ねた代名詞のことです．関係代名詞には，関係代名詞によって導かれる形容詞節を先行詞(名詞)に結び付ける限定用法と，先行詞についての説明を追加する継続用法(非限定用法)があります．関係代名詞は，学術論文中においてもよく用いられます．

先行詞が人の場合で，関係代名詞節のなかで関係代名詞が主格の場合には "who" が，所有格の場合には "whose" が，目的格の場合には "whom" が用いられます．先行詞が事物の場合で，関係代名詞節のなかで関係代名詞が主格または目的格の場合には "which" が，所有格の場合には "of which" または "whose" が用いられます．関係代名詞の "that" は，先行詞が人あるいは事物の場合で，関係代名詞節のなかで関係代名詞が主格あるいは目的格の場合に用いることができます．

まずは，関係代名詞の限定用法の使用例を以下に示します．

> Lightning is an electric discharge whose path length is measured in kilometers.
> =Lightning is an electric discharge the path length of which is

measured in kilometers.

（＝Lightning is an electric discharge. Its path length is measured in kilometers.）

雷は，その長さが数 km にもなる放電です．

この例では，先行詞が "an electric discharge"（事物）で，関係代名詞節のなかで関係代名詞が所有格となっています．

Our goal is to produce a Web page in which an animation sequentially brings up pictures.

（＝Our goal is to produce a Web page. In the Web page, an animation sequentially brings up pictures.）

アニメーションが連続して写真を繰り出すようなウェブページの製作が私たちの目標です．

この例では，先行詞が "a Web page"（事物）で，関係代名詞が前置詞 "in" の目的語になっています．

先行詞に最上級の形容詞，序数詞，"the same"，"the only" などがついている場合には，関係代名詞として "that" が用いられます．

It is the only method that we can use for computing transient electromagnetic fields in a complex structure.

（＝It is the only method. We can use it for computing transient electromagnetic fields in a complex structure.）

それが，複雑な構造物内の過渡電磁界計算に使える唯一の方法です．

"the thing(s) which"（すること，するもの）を意味する "what" は，複合関係代名詞と呼ばれています．その使用例を以下に示します．

What you can see here is the cross-section of a power cable.

（＝The thing which you can see here is the cross-section of a power cable.

＝The thing is the cross-section of a power cable. You can see it here.）

こちらでご覧になれますのは電力ケーブルの断面です．

次に，関係代名詞の継続用法の使用例を以下に示します．この用法は，長い文を読みやすくするのに役立ちます．

A moving-coil instrument, which measures only direct current, may be used in conjunction with a bridge rectifier circuit to provide an indication of alternating currents and voltages.

可動コイル形計器は，直流のみを計測する計器ですが，ブリッジ整流回路とつなぐことによって，交流電流や交流電圧の指示に用いられます．

Large coal-fired and nuclear power plants, which are the most powerful generating units, are normally designed to operate at full or nearly full capacity.

大規模石炭火力発電所と原子力発電所は，最も強大な発電ユニットで，通常，それらはフル稼働かほぼフル稼働するように設計されています．

The bipolar junction transistor consists of three regions of semiconductor material. One type is called a p-n-p transistor, in which two regions of p-type material sandwich a very thin layer of n-type material.

バイポーラ接合トランジスタは半導体物質の三つの領域からなります．第1のタイプはp-n-pトランジスタと呼ばれており，そのなかのp形物質の2領域がn形物質の非常に薄い層を挟んでいます．

継続用法の関係代名詞の前には，通常，カンマが置かれます．関係代名詞節が文の途中に入る場合には，その前後にカンマを置きます．"that"は継続用法では用いられません．最後の例文の場所を表す"in which"は関係副詞"where"に置き換えることができます．"in which"や"at which"が時を表す場合には，それらは関係副詞"when"に置き換えることができます．

理由を表す関係副詞"why"の先行詞は"the reason"と決まっています．この使用例を以下に示します．

> The reason why a major revision is needed is understandable.
>
> 大幅改正が必要とされている理由が理解できます．

方法を表す関係副詞 "how" には先行詞が含まれており，"the way how" とはいいません．この使用例を以下に示します．

> This is how we derived this expression.
>
> このようにして，私たちはこの式を導きました．

〈補足〉 これはやや強い表現で，「新しい方法で導いた」という意味が込められています．"This is the way (in which) we derived this expression." もほぼ同じ意味ですが，「既存の方法を適用して導いた」というニュアンスをもちます．"This method was used to derive the expression." と書くと，さらに控えめな意味になります．

1.7 前置詞と群前置詞の使い方
How to Use Prepositions and Group Prepositions

本節では，学術論文中での前置詞と群前置詞の使い方について説明します．前置詞は，名詞，代名詞，名詞相当語句の前に置かれて，形容詞句や副詞句を作ります．"apply...to〜"（...を〜に適用する）の例からもわかるように，その前にくる動詞（上の例では apply），形容詞あるいは名詞によって相応しい前置詞が決まることが多いため，論文中で頻繁に使われるものについては覚えておくのが便利です．群前置詞は，"because of" や "instead of" のように，2語以上がまとまって一つの前置詞の働きをします．

1.7.1 前置詞の使い方 *How to Use Prepositions*

前置詞は，名詞，代名詞，名詞相当語句の前に置かれて，形容詞句や副詞句を作ります．その前に置かれる動詞，形容詞あるいは名詞によって相応しい前置詞が決まることが多いため，下記に示すような頻繁に用いられるもの

については覚えておくのが便利です．

「適用する」という意味の動詞 "apply" は，下記のように前置詞 "to" をとります．

> We have applied the finite-difference time-domain (FDTD) method to computing lightning electromagnetic pulses (LEMPs).
>
> 時間領域有限差分(FDTD)法を雷電磁界パルス(LEMP)の計算に適用しました．

なお，前置詞 "to" の目的語を動名詞(gerund)の "computing" ではなく，"a lightning electromagnetic-pulse computation" のように名詞に変更しても同じ意味になります．また，"apply" の名詞形(application)を用いる場合にも，次のように前置詞として "to" を選びます．"Application of the FDTD method to a LEMP computation"（FDTD 法の LEMP 計算への適用）．ここで，"LEMP" の不定冠詞が "an" ではなく "a" となっているのは，"LEMP" が「エルイーエムピー」ではなく，「レンプ」と発音されるためです．同様に，"local area network" の頭字語である "LAN" は「ラン」と発音されるため，不定冠詞は "a" となります．一方，"light emitting diode" の頭字語である "LED" は「エルイーディ」と発音されるため，不定冠詞は "an" となります．学術論文中では，あまり用いられませんが，「…に応募する」という意味で "apply" を用いる場合には，前置詞として "for" を選ぶ必要があります．

「接続する」という意味の動詞 "connect" は，下記のように前置詞 "to" または "with" をとります．

> A 10-μF capacitor is connected to an 80-Ω resistor.
>
> 10 μF のキャパシタが 80 Ω の抵抗器に接続されています．

「直列に」接続されていることを示す場合には，"to" の代りに "in series with" を，「並列に」接続されていることを示す場合には，"in parallel with" を用います．

「比較する」という意味の動詞 "compare" も，下記のように前置詞 "with" または "to" をとります．

> We have compared our computed results with the corresponding

> measured values.
>
> 私たちの計算結果を対応する測定値と比較しました.

なお，比較を表す "compared with" や "in comparison with" を下記のように形容詞と一緒に用いる場合には，形容詞は比較級にはせず原級とします.

> Our computed results are reliable compared with their reported results.
> （＝Our computed results are more reliable than their reported results.）
>
> 私たちの計算結果は彼らの報告結果よりも信頼できます.

学術論文中では，あまり使われませんが，「...に例える」という意味で "compare" を用いる場合には，"to" を選ぶ必要があります（例：Life is compared to a voyage.）.

「一致する」という意味の動詞 "agree" は，下記のように前置詞 "with" をとります.

> Our results agree well with the theoretical predictions.
>
> 私たちの計算結果は理論的に予測したものによく一致しています.

名詞形（agreement）を用いる場合にも，"with" を選びます.

「似ている」という意味の形容詞 "similar" は，下記のように前置詞 "to" をとります.

> We have adopted an approach similar to the method of moments.
>
> モーメント法に似た手法を採用しました.

この例文の "similar to" を前置詞 "like"（...に似た，...のような）に置き換えてもほとんど同じ意味になりますが，口語的な感じになります.

「同じの」という意味の形容詞 "same" は，下記のように前置詞 "as" をとります.

> The surface area of this country is almost the same as the United Kingdom.
>
> この国の面積はイギリスとほぼ同じです．

"the same..." の後に節がくる場合には，次のように，"as" の代わりに "that" を用いることもできます．"We will adopt the same method that they used."

「同一の」という意味の形容詞 "identical" は，下記のように前置詞 "to" または "with" をとります．

> This voltage waveform computed using the FDTD method is almost identical to the corresponding measurement.
>
> FDTD 法を用いて計算したこの電圧波形は対応する測定波形とほぼ一致しています．

「依存する(左右される)」という意味の形容詞 "dependent" は，下記のように前置詞 "on" または "upon" をとります．

> The conductivity of this semiconductor is highly dependent on temperature.
>
> この半導体の導電率は温度に強く依存します．

「依存しない(独立した)」という意味の形容詞 "independent" は，下記のように前置詞 "of" をとります．

> The conductivity of this alloy is nearly independent of temperature.
>
> この合金の導電率は温度にほとんど依存しません．

「異なる」という意味の形容詞 "different" は，下記のように前置詞 "from" をとります．

> Electrical properties of this new matetrial are different from those of any conventional material.
>
> この新素材の電気的特性はいかなる従来素材の特性とも異なっています．

「できる」という意味の形容詞 "capable" は，下記のように前置詞 "of" をとります．

> This thermometer is capable of measuring body temperature without physical contact.
>
> この温度計は体温を非接触で測定することができます．

「原因（理由）」という意味の名詞 "reason" は，下記のように前置詞 "for" をとります．

> The reason for this discrepancy has not yet been understood.
>
> この不一致の原因はまだ明らかになっていません．

手段や方法を表す前置詞として，下記のように "with" がよく用いられています．

> We observed dislocations in a silicon phototransistor with a scanning electron microscope (SEM).
>
> 私たちは，走査形電子顕微鏡(SEM)を用いてシリコン光トランジスタの転位を観測しました．

"with" の代りに，前置詞のように使うことが許されている分詞 "using" も頻繁に用いられています．人や計算機「による」ことを表す場合には，下記のように "by" が用いられます．

> He performed this integral using separation of variables.

> 彼は，この積分を変数分離法を用いて行いました．
>
> The integral was evaluated by computer using the trapezoidal rule.
>
> その積分は台形則を用いて計算機により行われました．

プログラミング言語「による」ことを表す場合には，下記のように前置詞 "in" が用いられます．

> This program is written in C++.
>
> このプログラムはC++で記述されています．

「…の」増加あるいは「…の」減少を表す場合，下記のように "in" が用いられます．

> With increasing temperature, most conductors increase in resistance and insulators decrease in resistance.
>
> 温度が上昇するにつれて，ほとんどの導体の抵抗は上昇し，絶縁体の抵抗は低下します．

「時間とともに減少する．」，「高さとともに減少する．」は，それぞれ，"It decreases with time.", "It decreases with (increasing) height." で表されるように，"with" が用いられます．

1.7.2　群前置詞の使い方　*How to Use Group Prepositions*

学術論文中でよく使われる群動詞は多くはないので，それらの使い方を覚えてしまうのが便利です．

原因や理由を表す群前置詞として，"because of", "owing to", "on account of", "due to" がよく用いられています．"due to" 以外は下記のように副詞句として用いられます．

> The common cloud-to-ground lightning has been studied more extensively than other forms of lightning because of its practical

> interest.
>
> 実際的な関心から，通常の対地雷は，ほかの形態の雷より広く研究されてきました．

"due to"は副詞句としては用いられず(したがって，上記の例文において"because of"の代りに"due to"を用いるのは間違いです)，形容詞句(限定用法および叙述用法)として用いられます．下記は，"due to"が前の名詞(increase)を修飾する限定用法で用いられている例と主格補語として叙述用法で用いられている例です．

> The increase in resistance of a conductor due to skin effect cannot be mitigated.
>
> 表皮効果に起因した導体の抵抗上昇は軽減できません．
>
> The increase in resistance of a conductor at higher frequencies is due to skin effect.
>
> 高周波数における導体の抵抗上昇は表皮効果が原因です．

根拠を表す群前置詞として，"on the basis of"が用いられています．この群前置詞を副詞句として用いた例を以下に示します．

> On the basis of Maxwell's equations, we have derived a new expression.
>
> マクスウェル方程式に基づいて，新しい式を導出しました．

最近では，"on the basis of"の代りに，"based on"が副詞句として用いられる場合もあります．もちろん，"due to"の使用例と同じように，"based on"は形容詞句(たとえば，"This new expression based on Maxwell's equations is..."のような限定用法および"This new expression is based on Maxwell's equations"のような叙述用法)として，より頻繁に用いられています．

目的を表す群前置詞として，"for the purpose of"や"for the sake of"があ

ります.

> We have developed this instrument for the purpose of measuring lightning currents.
>
> 私たちは，雷電流を測定するために，この計器を開発しました．

ただし，前置詞 "for" のみでも十分「目的」を表すことができます．また，"We have developed this instrument (in order) to measure lightning currents." のように，不定詞(infinitive)を用いた文で同じ意味を表すこともできます．

手段や方法を表す群前置詞として，"by means of"，"by use of"，"with the help of" が用いられることがあります．

> We observed dislocations in a silicon phototransistor by means of a SEM.
>
> 私たちは，SEM を用いてシリコン光トランジスタの転位を観測しました．

しかし，これらの群前置詞とほとんど同じ意味を表す前置詞 "with" あるいは前置詞のように使うことが許されている分詞 "using" がより頻繁に用いられています．なお，"by using" という表現は，ほとんど用いられていません．

代替や代理を表す群前置詞として，"instead of"，"in place of"，"on behalf of" がよく用いられています．

> I purchased this new computer instead of that.
>
> 私は，あの計算機の代りに，この新しい計算機を買いました．

この文では，"instead of" の代りに，"in place of" を用いても同じ意味になります．"in place of" は人が(この群前置詞の)目的語の場合にも用いられます．"on behalf of"（代理で，代表して）は，人または人の団体が目的語の場合にしか用いられません．

そのほかのよく使われる群前置詞の使用例を以下に示します．

In general, we classify materials according to their electrical properties into three types: conductors, semiconductors, and insulators.

一般に，私たちは物質をその電気的特性に従って，導体，半導体，そして絶縁体の3種類に分類します．

Your paper should be prepared in accordance with the guide for authors.

論文は，投稿規定に従って作成されなければなりません．

Contrary to my prediction, I was not able to develop a new efficient method.

予想に反して，新しい効率的手法を開発することができませんでした．

〈補足〉「…と対照的に」は "in contrast to …" と書きます．

We compared these projects in terms of cost.

私たちは，これらの計画をコストの点から比較しました．

Light-emitting diodes now have everything they need to dominate lighting, except for a sufficiently low price.

発光ダイオードは，今や照明を独占するために必要な，十分安い価格以外のすべてをもっています．

In addition to the desirable effect of inducing an electromotive force in the coil, the alternating flux induces undesirable voltages in the iron core.

コイルに起電力を誘起するという望ましい効果に加えて，その交流磁

束は鉄芯に望ましくない電圧を誘起します．

This system is used here in spite of its low efficiency.

低効率にもかかわらず，ここでは，このシステムが用いられています．

Regardless of age, everyone can apply for it.

年齢に関係なく，誰でもそれに応募できます．

〈補足〉 "regardless of" の代りに，"irrespective of" を用いても同じ意味になります．

A 10-µF capacitor is connected in parallel with an 80-Ω resistor.

10 µF のキャパシタが 80 Ω の抵抗器に並列に接続されています．

〈補足〉 「…に直列に」は "in series with …" と書きます．

Capacitive reactance is in inverse proportion to frequency.

容量性リアクタンスは周波数に反比例します．

〈補足〉 "in inverse proportion to" よりも "inversely proportional to" という表現がより一般的です．

1.8 ハイフン，ダッシュ，コロン，セミコロンの使い方
How to Use a Hyphen, Dash, Colon, and Semicolon

本節では，英語の学術論文中で使われるハイフン（-），ダッシュ（—），コロン（：），セミコロン（；）の使い方について説明します．

1.8.1 ハイフンの使い方　*How to Use a Hyphen*

ハイフン(-)は，別々の単語をつないで複合語を作るのに用いられます．特に，下記のように，複合語がほかの名詞を修飾する場合(形容詞の限定用法的に使用される場合)によく用いられます．

High-voltage engineering

高電圧(の)工学

Power-line-communication signal

電力線搬送通信(の)信号

Semiconducting-layer conductivity

半導電層の導電率

ただし，複合語が形容詞の叙述用法的に使用される場合や名詞的あるいは副詞的に使用される場合には，ハイフンは不要です．また，ハイフンがなくても誤解や混乱が生じない場合には，複合語が形容詞の限定用法的に使用される場合においても，ハイフンが省略されることがあります．たとえば，"power-line-communication signal" が，"power-line communication signal" または "power line communication signal" と表記されたり，"an 80-Ω tesistar" が "an 80 Ω resistor" と表記されることもあります．

"semiconducting" も，もともとは接頭辞 "semi-" と現在分詞 "conducting" をつなげたものですが，現在ではハイフンを入れずに一体化して用いられています．

2語から構成される99未満の数はハイフンでつなげて表記されます．たとえば，56は "fifty-six" と書かれます．

1.8.2 ダッシュの使い方　*How to Use a Dash*

ダッシュ(—)は，下記のように，説明を追加したり，詳しく述べたりする場合に用いられます．括弧の使い方とよく似ています．

> Iron losses are of two types—hysteresis loss and eddy current loss.
>
> 鉄損には2種類(ヒステリシス損と渦電流損)あります．
>
> The rearrangement of charge occurs very rapidly—within microseconds across a 0.1-m conductor—so it is generally safe to assume that the electric field inside the conductor is equal to zero.
>
> 電荷の再配列は非常に速やかに(0.1 m の導体で数マイクロ秒以内に)起こるため，導体内の電界は零に等しいと考えても，ほとんどの場合，差し支えありません．

上述のダッシュを "em dash" といいます．この半分程度の長さの記号(–)で表されるダッシュは "en dash" といいます．このダッシュは，下記のように，数値の範囲を表すのに用いられます．

> pp. 100–150
>
> 100 ページから 150 ページまで
>
> A voltage of 100–150 V
>
> 100 から 150 V の電圧

ページの範囲を表す場合には，問題ありませんが，電圧の範囲などを表す場合には，ダッシュ記号がマイナス記号と誤解される可能性があります．このような場合には，ダッシュ記号の代りに前置詞 "to" を用いて，"A voltage of 100 to 150 V" と表記します．

なお，x' は "x dash" と読まれる場合もありますが，通常は，"x prime" と読まれます．

1.8.3 コロンの使い方　*How to Use a Colon*

コロン(：)は，コロンの前の内容を詳しく説明したり，細目を列挙したり，具体例をあげたり，数式を導入したり，引用文を挿入する場合に用いられます．

細目を列挙する場合のコロンの使用例を以下に示します．

> In general, we classify materials according to their electrical properties into three types: conductors, semiconductors, and insulators (or dielectrics).
>
> 一般に，私たちは物質をその電気的特性に従って，導体，半導体，そして絶縁体（または誘電体）の3種類に分類します．

このように，細目を列挙する場合には，コロンに続く単語の頭文字は小文字にします．

具体例をあげる場合の使用例を以下に示します．

> Electric fields obey the principle of superposition: The electric field due to a set of sources is the vector sum of the individual electric fields due to each source.
>
> 電界は重ね合せの原理に従います．1組の源（電荷）による電界は，各源による個々の電界のベクトル和です．

このように，コロンの後ろに文が続く場合には，ピリオドに近い役割を果たしていると考え，コロンに続く文における最初の単語の頭文字を大文字にします．

数式を導入する場合の使用例を以下に示します．

> Vector subtraction can be defined in terms of vector addition in the following way:
> $$A - B = A + (-B),$$
> where $-B$ is the negative of vector B.
>
> ベクトル減法は次のようにベクトル加法として定義することができます．
> $$A - B = A + (-B),$$
> ここで，$-B$ はベクトル B の負であるとします．

1.8.4　セミコロンの使い方　*How to Use a Semicolon*

セミコロンは，下記のように，なんらかの相互関係がある二つの（等位の）文をつなげる場合によく用いられます．カンマ（,）より強く，ピリオド（.）よりは弱い句読点（punctuation）と考えてよいでしょう．

> In polar latitudes you may see the aurora; in Japan you are more likely to see a rainbow. Both of these phenomena are explained by the laws of electromagnetism.
>
> 極緯度においてはオーロラを見ることがあるかもしれません．日本では虹を見る可能性が高いでしょう．どちらの現象も電気磁気の法則で説明されます．

前後関係を明確にするために，下記のように，セミコロンの後に等位接続詞（この例では but）を挿入する場合もあります．

> Many orthogonal coordinate systems exist; but we are concerned only with the following three systems:
> - Cartesian;
> - Cylindrical; and
> - Spherical.
>
> 多くの直交座標系があります．しかし，私たちは次の三つの座標系のみに関心があります．
> - デカルト座標
> - 円筒座標
> - 球座標

箇条書きの左端の黒丸（•）は "bullet" といいます．箇条書きリストの区切りのセミコロンは，カンマで置き換えることもできますし，セミコロンやカンマなしとすることもできます．ただし，同一論文においては，これらのうちの一つを統一して用いなければなりません．

セミコロンは，下記のように，同格の語句を区切る場合にも用いられます．これは，論文中でのハーバード方式により文献を引用している例です．なお，この例文でのセミコロンをカンマで置き換えると，著者名と出版年の関係が

わかりにくくなりますので，セミコロンの使用が望ましいです．

> The method of moments in the time domain (Van Baricum and Miller, 1972; Miller et al., 1973) is widely used in analyzing electromagnetic responses of thin-wire structures.
>
> 時間領域のモーメント法(Van Baricum and Miller, 1972; Miller et al., 1973)は，細線構造の電磁界応答を解析する際に広く用いられています．

1.9 論文投稿前の文法と剽窃チェックの仕方
How to Check Grammar and Plagiarism Before Paper Submission

1.9.1 文法とつづりのチェックの仕方
How to Check Grammar and Spelling

　学術論文の目的は，その分野の専門家に新しい知見を伝えることにあります．しかし，文法や単語のつづりが誤っていると，査読者に重要な成果を正しく理解してもらえなかったり，読者に最後まで読んでもらえなかったりする可能性があります．英文の校正ツールを利用しながら論文を書いたり，投稿前に英文校正ツールを利用して英文チェックを行うことで，英文法やつづりの誤りを少なくすることができます．

　文書作成ソフトとして Microsoft 社の Word を用いて英語論文を書く場合には，「メニュー」から「校閲」機能を選び，さらに「スペルチェックと文章校正」機能を選択すれば，つづりの誤りとその修正案候補が示されます．この機能を活用すれば，つづりの誤りを探し出す手間が省け，誤記の少ない英語論文を執筆することができると思います．

　さらに役立つ英文校正ツールも複数存在しています．例えば，Grammarly 社の Grammarly が挙げられます．このツールは，文法やつづりの誤りのチェックとそれらの修正案候補の提案だけではなく，より相応しい英文表現の提案もしてくれます．このため，高度な英文校正をリアルタイムで受けながら，効率的に英語論文の執筆を行うことができます．また，英語論文執筆時だけではなく，英文で電子メールを書いているときにも，リア

ルタイムで誤りが指摘され，その修正案候補が表示されたり，より適した表現が提案されたりするので，とても便利です．機能に制限はありますが，無料版もあります．

英文校正ツールではありませんが，Google 社の機械翻訳ツール Google Translate は，日本語や英語以外で書かれた文書内容を日本語や英語に翻訳する際に役立ちます．ただし，一般公開されていない文書，特に査読を依頼された論文原稿など機密性の高い文書の翻訳はこのツールで行うべきではありません．Google Translate の翻訳の正確さは，言語ペア(例えば，ドイツ語-英語，ドイツ語-日本語など)や元の文書の記述レベルにも依存します．

日本語の文書を，Google Translate で英文に全翻訳するとどのようになるか見てみましょう．例えば，この節の第一パラグラフの日本語文を Google Translate で英文に翻訳すると，次のような文章が出力されます．

> The purpose of academic papers is to convey new knowledge to experts in the field. However, incorrect grammar and spelling of words can prevent reviewers from correctly understanding important results and readers from reading to the end. You can reduce errors in English grammar and spelling by using English proofreading tools when writing your thesis, and by using English proofreading tools to check your English before submission.

"thesis" を "paper" に変更した方が良いですが，元の日本語文書の内容を誤りなく翻訳できています．この英訳された文章を Google Translate で日本語に翻訳すると，次のような文章が出力されます．

> 学術論文の目的は，その分野の専門家に新しい知識を伝えることです．ただし，文法や単語のスペルが間違っていると，査読者が重要な結果を正しく理解し，読者が最後まで読むことができなくなる可能性があります．論文執筆時に英語校正ツールを使用したり，提出前に英語校正ツールを使用して英語をチェックしたりすることで，英語の文法やスペルの間違いを減らすことができます．

この日本語の文章からも，英語に機械翻訳された文章に誤りがないことが確認されます．

1.1.6 の「緒言」の第一パラグラフの日本語訳を，Google Translate で英語に翻訳すると，次のような英文が出力されます．

第 1 章　英語論文執筆法

> Recently, power line communication (PLC) has attracted a lot of attention. Power lines and power cables are used in PLC systems for data communication at frequencies up to 30 MHz. Power cables usually have a semiconducting layer between the center conductor and the insulating layer and between the insulating layer and the sheath conductor. These semiconducting layers suppress the occurrence of dielectric breakdown from unexpected protrusions on the surface of the central conductor and the sheath conductor, and also accommodate the thermal expansion of the insulating layer. Power cables are not designed to efficiently transmit PLC signals, so these can be significantly attenuated on power cables due to the presence of semi-conductive layers.

この英文も元の日本語の文章の内容を誤りなく翻訳できています．この英訳された文章を Google Translate で日本語に翻訳すると，次のような文章が出力されます．

> 近年，電力線通信(PLC)が注目を集めています．電力線と電力ケーブルは，PLC システムで最大 30 MHz の周波数でのデータ通信に使用されます．電力ケーブルは通常，中心導体と絶縁層の間，および絶縁層とシース導体の間に半導電層を備えています．これらの半導電層は，中心導体やシース導体の表面の予期せぬ突起による絶縁破壊の発生を抑制するとともに，絶縁層の熱膨張を吸収します．電源ケーブルは PLC 信号を効率的に送信するように設計されていないため，半導電層の存在により PLC 信号が大幅に減衰する可能性があります．

この日本語の文章からも，英語に機械翻訳された文章に誤りがないことが確認されます．

上記のいずれの場合も，元の日本語文章の各文が比較的単純な構造ですので，誤りなく英訳されたといえます．複雑で長い文では誤訳が発生することがあります．

なお，論文原稿の内容は，出版されるまでは，重要な機密情報ですので，機械翻訳の利用については，指導教授や上司とよく相談して決めてください．また，機械翻訳の利用は，英語論文執筆作業の効率化に役立つ可能性がある一方で，自身の英語力の向上を妨げてしまう可能性も高いという負の側面も

あります．

1.9.2　剽窃チェックの仕方　*How to Check Plagiarism*

　著者自身の過去の論文も含め他の論文で発表された研究成果や他の著作物の文章を，それらの出典を明記せずに自身の論文などで使用することを剽窃あるいは盗用といいます．剽窃を行うと，著者(共著者も含む)は，剽窃が明らかになった以降の一定期間，その出版社や学会の雑誌への投稿が禁止されるという罰則を受けます．また，著者自身やその所属組織の評判や信用も低下させてしまいます．

　意図的に剽窃を行う人はほとんどいないと思いますが，意図せずに剽窃が発生する場合があります．例えば，参考文献情報の記載を誤り，該当する文献を引用していなかったりする場合などです．また，生成的人工知能(生成AI)や機械翻訳を利用した場合，生成された文章や翻訳された文章が，第三者が過去に発表した著作物のものと似通ったものになる可能性もあり得るのに，その出典は明らかではないため，知らずのうちに著作権を侵害したり，剽窃行為を行ってしまう恐れもあります．剽窃とは異なりますが，生成AIに査読コメントを生成してもらうことは他人の未発表の成果という機密情報をデータベースに残してしまう危険性がありますし，自ら専門家であることを否定する行為でもありますので，避けましょう．

　意図しない剽窃を未然に防ぐためには，例えば，オンラインの剽窃検知ツールを用いて，過去の論文との類似度あるいは重複度のチェックを行う必要があります．例えば，iParadigms社の剽窃検知ツールiThenticateに投稿前の論文原稿をアップロードすると，著者自身の過去の論文を含め，データベースに収録されている国内外の雑誌論文や学会発表論文との重複率が計算され，過去の各論文との重複部分，重複率および総重複率などが出力されます．この結果から，見落としていた文献に気付き，それを参考文献に追加することができます．なお，通常は，著者名，キーワード，参考文献，著者略歴なども含めて重複率が算出されますし，研究者はある大きな課題に対して，研究を少しずつ進め，ある程度まとまった新しい知見を論文にまとめて順番に発表していくため，著者本人の過去の論文との重複がある程度発生するのが普通です．このため，著者本人の投稿前の論文原稿と著者本人の過去の各論文との重複度がそれぞれ数パーセントであっても，総重複率が30％程度に積み上がることもあります．この総重複率が投稿予定の雑誌が決めている重複率の許容値を超えていれば，投稿しても著者にすぐに差し戻されますので，例えば，論文の緒言や研究方法などの重複部分を簡略化し，過去論文の引用

で置き換えるなどして，総重複率を許容範囲内に抑えてから投稿する必要があります．

第 2 章

英語論文口頭発表法

Techniques for Presenting Successful Scientific Papers in English

2.1　発表用スライドの作成法と発表法
2.2　グラフや表の説明法
2.3　数と数式の読み方
2.4　質問への対処法
2.5　国際会議での座長のことば
2.6　発表用ポスターの作成法と発表法
2.7　オンライン発表の心得とオンデマンド発表資料の作成

2.1 発表用スライドの作成法と発表法
How to Prepare Slides for an Oral Presentation and How to Speak with the Slides

本節では,主として口頭発表において最近もっぱら用いられているパワーポイントスライドの作成法とそれを用いた発表法について説明します.

2.1.1 スライドの構成 *How to Organize Slides*

発表時間 15 分,質疑応答時間 5 分の講演を例にして説明します.15 分間の発表ですと,スライドは 10 枚程度が適当であると思います.平均してスライド 1 枚に対して,1 分 30 秒程度は必要です.かぎられた枚数のスライドに論文のすべての内容を載せることはできませんので,論文の最も重要な部分のみに絞り,大きな文字,大きな図を用いてスライドを作成しましょう.

表 2.1 発表用スライドの基本構成(発表時間が 15 分の場合の例)

順番	項目	スライド枚数の目安	発表時間の目安(分)
1	発表論文題目,著者名,所属 (Title, authors' names, affiliations)	1	0.5
2	発表内容の概要(Outline or contents) これから発表する内容の順序とアウトラインを示します	1	1
3	研究の背景と目的(Background and objective) これから発表する研究を実施するきっかけとなった背景と研究目的を示します	1	2
4	研究方法(Method, theory or model) 研究方法,理論やモデルについて示します	1	2
5	研究結果(Results) 最も見せたい結果を,できるだけ大きい図にして,わかりやすく説明します	2〜3	4
6	考察(Discussion) 得られた結果の妥当性や物理的根拠について論じます	2〜3	4
7	まとめ(Summary or conclusions) 研究成果の最も大事な部分を箇条書きにして,示します	1	1.5

質疑応答の際に，「何ページ目のスライドをもう一度見せてください．」などといわれることが多いので，各スライドにページ番号をつけておくと役立ちます．スライドは，オーバーヘッドプロジェクタ用の透明シートの名残りで今もビューグラフ(viewgraph)と呼ばれることもあります．作成したスライドの電子ファイルは，USB メモリスティック(USB plug-in memory stick)に保存して会場に持参しましょう．

表 2.1 に発表用スライドの基本構成を示します．2.1.2 から 2.1.8 では，15 分の発表時間の講演用に作成したスライドを例に，発表の開始から終了までの流れを説明します．

2.1.2 発表の開始　*How to Start a Presentation*

座長による次のような論文題目，著者名，著者の所属についての紹介を受けて，発表を開始します．

> The title of the next paper is "Influence of semiconducting layers on propagation of power line communication signals in power cables". The authors are Ms. Hanako Denki and Professor Taro Denshi of the University of Southern Kyoto, Japan. The paper will be presented by Ms. Hanko Denki.
>
> 次の論文のタイトルは，「電力ケーブル上での電力線搬送通信信号の伝搬に与える半導電層の影響」です．著者は，南京都大学の電気花子さんと電子太郎教授です．ご発表は，電気花子さんにより行われます．

これを受けて，たとえば図 2.1 のようなスライドをスクリーンに表示し，次のように発表を開始します．

> Thank you, Mr. (または Ms.) chairperson for introducing me. Good morning (または Good afternoon). My name is Hanako Denki. I am a Ph.D. student of the University of Southern Kyoto. Today, I will be talking about influence of semiconducting layers on propagation of power line communication signals in power cables.
>
> 座長様，ご紹介ありがとうございます．おはようございます(こんにちは)．電気花子と申します．南京都大学の博士課程の学生です．今日

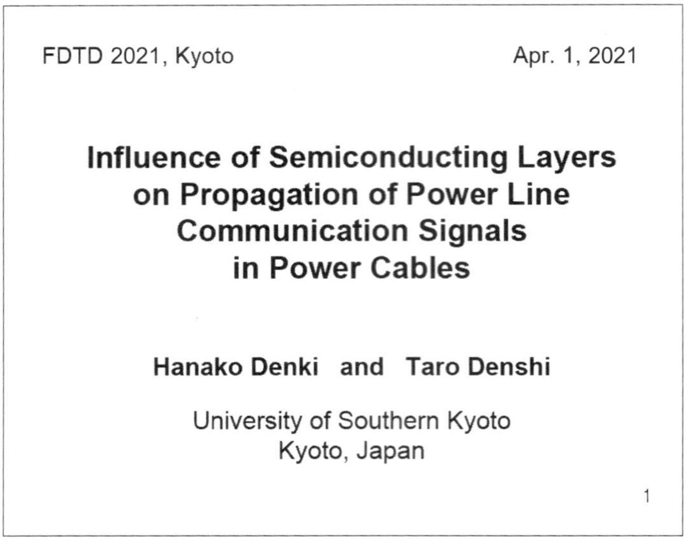

第2章　英語論文口頭発表法

> FDTD 2021, Kyoto　　　　　　　　　　Apr. 1, 2021
>
> **Influence of Semiconducting Layers on Propagation of Power Line Communication Signals in Power Cables**
>
> Hanako Denki　and　Taro Denshi
>
> University of Southern Kyoto
> Kyoto, Japan
>
> 1

図2.1　発表論文題目，著者名，所属を示したスライド

は，電力ケーブル上での電力線搬送通信信号の伝搬に与える半導電層の影響について発表いたします．

〈補足〉"chairman"というよりは，"gender-neutral word"である"chairperson"を用いましょう．発表では，1人称の"I"（発表者本人），"we"（発表者と共著者あるいは発表者と聴講者），および2人称の"you"（聴講者）を積極的に使いましょう．

招待講演の場合には，自己紹介の後に，招待してくださった方や組織に次のようなお礼を述べましょう．"I would like to thank Professor Jane Doe of the University of Southeastern America for inviting me to talk here today. I am very happy to be able to talk to you."

自分の声が会場の皆さんに聞こえているか否かを確認する場合には，次のようにいうとよいでしょう．"Can you hear me at the back?"

マイク（またはレーザポインタ）の調子が悪いときには，会場係の方に次のように伝えましょう．"This microphone（または laser pointer）is not working well."

また，コンピュータの調子が悪いときには，会場係の方に次のように伝えましょう．"There is something wrong with this computer."

セッション開始前に，発表に用いるパワーポイントスライドの動作確認をする場合には，会場係の方に次のようにいいましょう．"I am supposed to give a presentation in the next session. May I check if my PowerPoint file will work properly on this PC?"

2.1.3 発表概要の説明
How to Show the Outline or Contents

次は，発表概要の説明を行います．発表概要を最初に説明しておくことは，以降の発表内容を的確に理解してもらうのに役立ちます．たとえば図 2.2 のようなスライドをスクリーンに表示し，次のように説明を行います．

This slide shows the outline of my presentation. First, I would like to talk about the background and the objective of this work. Second, I will show the model of a power cable with semiconducting layers for analyzing the propagation characteristics of power line communication signals using the finite-difference time-domain method. Then, I will show analyzed results that include the dependence of propagation characteristics of PLC signals on the semiconducting-layer conductivity of the power cable. After that, I will discuss the attenuation and dispersion of a PLC signal, and its propagation velocity. Finally, I will summarize my presentation.

こちらに示しておりますのが私の発表のアウトラインです．まず，本研究の背景と目的についてお話します．次に，時間領域有限差分 (FDTD) 法を用いた電力線搬送通信 (PLC) 信号の伝搬特性を解析するための，半導電層を含む電力ケーブルのモデルを示します．その後，PLC

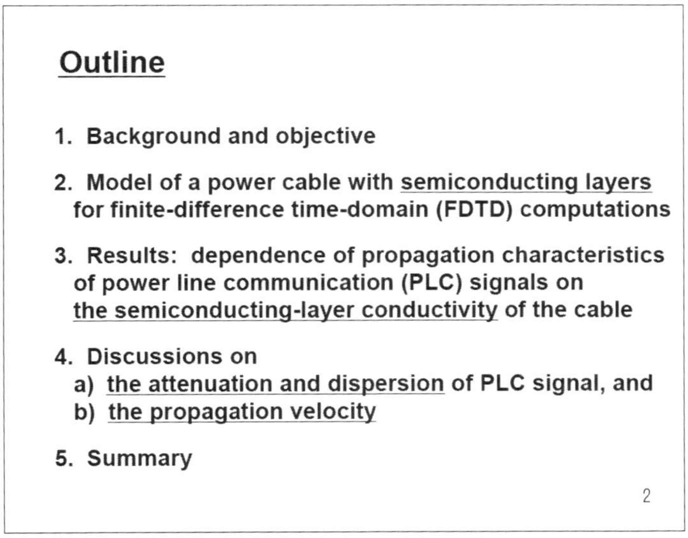

Outline

1. Background and objective

2. Model of a power cable with <u>semiconducting layers</u> for finite-difference time-domain (FDTD) computations

3. Results: dependence of propagation characteristics of power line communication (PLC) signals on <u>the semiconducting-layer conductivity</u> of the cable

4. Discussions on
   a) <u>the attenuation and dispersion</u> of PLC signal, and
   b) <u>the propagation velocity</u>

5. Summary

図 2.2 発表内容の概要を示したスライド

> 信号伝搬特性の電力ケーブル半導電層導電率への依存性を含む解析結果を示します．その後，PLC 信号の減衰および変わい，伝搬速度についての考察を行います．最後に，本発表のまとめを行います．

〈補足〉「最初に」は "first"，「最後に」は "finally"，それ以外は "then" でもよいでしょう．なお，"analyze"（解析する）の名詞形単数は "analysis"，複数形は "analyses"，形容詞形は "analytical" または "analytic" です．"analytical" と "analytic" の意味に差はほとんどありません．"electrical" と "electric" については，「放電」を "electric discharge" とも "electrical discharge" ともいえるように，両者の意味にほとんど差がない場合もあれば，そのどちらかが好まれる場合もあります．たとえば，"electrical engineering"（電気工学）とはいいますが，"electric engineering" とはいいません．

会場の皆さんに伝わらない可能性のある専門用語（jargon）の使用は避けましょう．また，"laser (light amplification by stimulated emission of radiation)" のようによく知られているもの以外の頭字語（acronym）を用いる場合には，最初に出た際に，もとの熟語や句を必ず示しましょう（finite-difference time-domain, FDTD; power line communication, PLC）．

なお，スライド枚数が 30 を超えるような長い講演の場合には，各セクションの説明が終わる際に，講演の現在地点と進行状況を確認する目的で，次のセクションの題目を目立たせた図 2.3 のようなスライドを示しましょう．

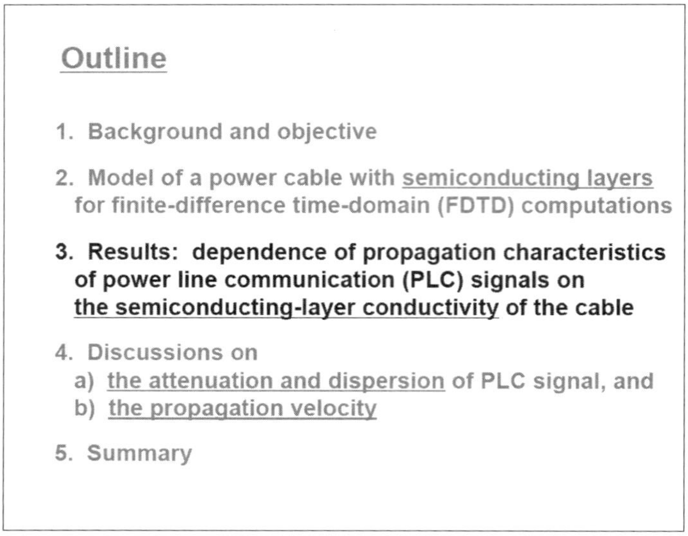

Outline

1. Background and objective

2. Model of a power cable with <u>semiconducting layers</u>
   for finite-difference time-domain (FDTD) computations

3. **Results: dependence of propagation characteristics
   of power line communication (PLC) signals on
   <u>the semiconducting-layer conductivity</u> of the cable**

4. Discussions on
   a) <u>the attenuation and dispersion</u> of PLC signal, and
   b) <u>the propagation velocity</u>

5. Summary

図 2.3　次に話すセクションを目立たせた発表概要を示すスライド

2.1.4 研究の背景と目的の説明
How to Explain the Background and Objective

次は，研究の背景と目的の説明を行います．どのような問題背景があり，どのような目的でこの研究を行ったのかについて簡潔に説明します．たとえば図 2.4 のようなスライドをスクリーンに表示し，次のように説明を行います．

Now, I would like to talk about the background and the objective of this work. Recently, considerable attention has been attracted to the PLC. The PLC systems use power lines and cables for data communication in a frequency range up to 30 megahertz. Within a power cable, semiconducting layers are incorporated to mitigate electrical breakdown from unexpected projections on the core and sheath conductor surfaces, and also to accommodate the thermal expansion of the insulating layer. Since power cables are not designed for effectively transmitting the PLC signals, they might attenuate significantly owing to the presence of semiconducting layers. The objective of this study is to perform FDTD simulations that reveal the influences of semiconducting layers on the PLC propagation characteristics.

それでは，これから本研究の背景と目的についてお話ししたいと思い

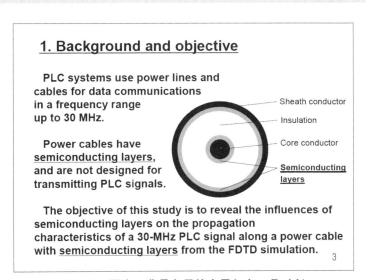

図 2.4　研究の背景と目的を示したスライド

ます．最近，電力線搬送通信PLCが大変注目されています．PLCシステムでは，30 MHzにいたるまでの周波数のデータ通信に電力線や電力ケーブルが使用されます．電力ケーブルには，ケーブル中心導体やシース導体表面の予想外の突起から発生する可能性のある絶縁破壊を抑制し，また絶縁層の熱膨張を吸収するために半導電層が組み込まれています．電力ケーブルはPLC信号を効果的に伝送するようには設計されてはいないため，PLC信号は半導電層の存在により著しく減衰する可能性があります．本研究の目的は，FDTDシミュレーションを実施し，PLC信号伝搬特性に与える半導電層の影響を明らかにすることです．

〈補足〉 本書では白黒でスライドを表示していますが，色使いを工夫したり，パワーポイントのアニメーション効果などを利用して，重要な部分を際立たせましょう．

「本研究の目的は…です．」は "The objective (purpose, aim または goal でもよい) of this study is…" といいます．

2.1.5 研究方法の説明　*How to Explain the Method*

次は，採用した実験方法，計算方法あるいは計算モデルについて簡潔に説明します．たとえば図2.5のようなスライドをスクリーンに表示し，次のように説明を行います．

What we can see here is the cross-section of the model of a power cable that we use in this work. The radius of the core conductor is 5 millimeters. The inner radius of the sheath conductor is 25 millimeters. Both the core and sheath conductors are perfectly conducting. The thickness of the insulating layer is 14 millimeters. This insulation thickness corresponds to a 66-kilovolt power cable. Between the core conductor and the insulating layer, and between the insulating layer and the sheath conductor, semiconducting layers are installed. The thickness of each semiconducting layer is 3 millimeters. We varied the layer conductivity in a range from 10^{-5} (ten to the minus five) to 10^5 (ten to the fifth) siemens per meter. We fixed the relative permittivity of the insulating and semiconducting layers at a typical value of 3.0.

　こちらは本研究で使用した電力ケーブルモデルの断面図です．中心導

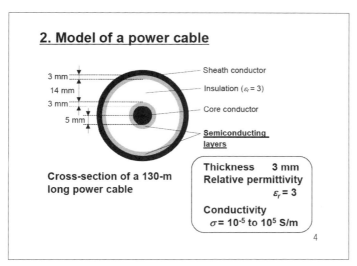

図 2.5 計算モデルを示したスライド

> 体の半径は 5 mm です．シース導体の内半径は 25 mm です．中心導体およびシース導体はともに完全導体です．絶縁層の厚みは 14 mm です．この絶縁厚は 66 kV クラスの電力ケーブルに相当します．中心導体と絶縁層の間および絶縁層とシース導体の間には半導電層が設けられています．各半導電層の厚みは 3 mm です．半導電層の導電率は 10^{-5} から 10^5 S/m の範囲に設定しました．絶縁層および半導電層の比誘電率は典型的な値である 3.0 にしました．

〈補足〉 "What we can see here is ..."「こちらは ... です．」はよく使う表現です．

10^5 S/m は "ten to the fifth siemens per meter" と読みます．"fifth" の後ろに "power" を挿入しても間違いではありませんが，省略されることがほとんどです．10^{-5} S/m は "ten to the minus five siemens per meter" と読みます．累乗の読み方については，2.3.3 で詳しく説明します．

2.1.6　結果の説明　*How to Explain the Results*

次は，最も見せたい実験結果あるいは計算結果について簡潔に説明します．たとえば図 2.6 および図 2.7 のようなスライドをスクリーンに表示し，次のように説明を行います．

> This slide shows FDTD-computed waveforms of a voltage pulse at distances of 20, 40, 60, 80, and 100 meters from the excitation point along

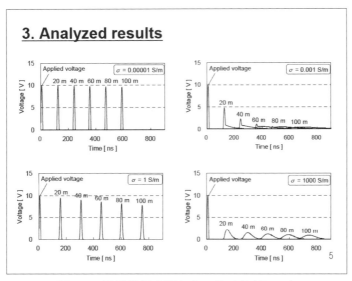

図 2.6　計算結果を示したスライド（その 1）

the power cable. The applied voltage is a positive half-sine wave of frequency 30 megahertz and magnitude 10 volts. The upper left figure shows the results for the semiconducting-layer conductivity of 10^{-5} (ten to the minus five) siemens per meter. The vertical axis represents the voltage between the core and sheath conductors in volts. The horizontal axis represents time in nanoseconds. The upper right figure shows the results computed for the semiconducting-layer conductivity of one millisiemen per meter. The lower left figure shows the results for the conductivity of one siemen per meter. The lower right figure shows the results for the conductivity of 1 000 siemens per meter.

From these results, you can see the magnitude of a voltage pulse decreases with increasing propagation distance. The dependence of the signal attenuation on the semiconducting-layer conductivity is not monotonic. The attenuation is significant at two points, when the semiconducting-layer conductivity equals one millisiemen per meter and 1 000 siemens per meter, while signals are unaffected when the semiconducting-layer conductivity is around one siemen per meter. Pulse width broadening or dispersion is also marked for semiconducting layers with high attenuation.

From these results, you can see the propagation velocity of the voltage pulse also depends on the semiconducting-layer conductivity.

The velocity is lowest when the semiconducting-layer conductivity is one siemen per meter.

　このスライドは，電圧印加点から 20，40，60，80 および 100 m 離れた点における電力ケーブル上の電圧パルスの FDTD 計算波形を示しています．印加電圧は周波数 30 MHz，振幅 10 V の正弦波の正半波です．上段左側の図は半導電層の導電率が 10^{-5} S/m の場合の計算結果を示しています．縦軸は，中心導体とシース導体間電圧をボルト単位で示しています．横軸は時間をナノ秒単位で示しています．上段右側の図は半導電層の導電率が 1 mS/m の場合の計算結果を示しています．下段左側の図は半導電層の導電率が 1 S/m の場合の計算結果を示しています．下段右側の図は半導電層の導電率が 1 000 S/m の場合の計算結果を示しています．

　これらの結果から，電圧パルスの振幅が伝搬に伴って減衰していることがわかると思います．この信号の減衰の半導電層の導電率への依存性は単調ではありません．半導電層の導電率が 1 mS/m および 1 000 S/m の 2 点において減衰は顕著となっています．一方，半導電層の導電率が 1 S/m の場合には，信号は影響を受けていません．減衰が著しくなる半導電層では，パルス幅の広がりあるいは分散もまた顕著となります．

　これらの結果から，伝搬速度も半導電層の導電率に依存していることに気づかれたと思います．半導電層の導電率が 1 S/m の場合には，ほかの導電率の場合に比べて伝搬速度が最も低くなっています．

〈**補足**〉 "You can see …"「… に気づかれたと思います．」または「ご覧のように … です．」はよく使う表現です．"The vertical (または horizontal) axis represents …"「縦軸 (または横軸) は … を示しています．」もよく使う表現です．スライドに示された図を理解するのに必要な時間を会場の方々に与えることにもなりますので，軸ラベルはできるかぎり読み上げましょう．

1 000 S/m を "one kilosiemen per meter" というのはまれで，ほとんどの場合，"one thousand siemens per meter" といいます．一方，"0.001 S/m" については，"zero point zero zero one siemen per meter" という場合は少なく，"one millisiemen per meter" というのが一般的です．

　Please take a look at the figure on the left. It shows the dependence of the magnitude of a 30-megahertz signal at distance 100 meters from the excitation point on the semiconducting-layer conductivity. The vertical

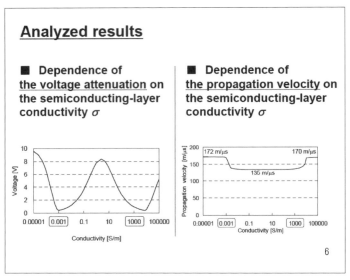

図 2.7　計算結果を示したスライド(その 2)

axis represents the voltage magnitude in volts. The horizontal axis represents the semiconducting-layer conductivity in siemens per meter. You can see the dependence of the signal attenuation on the seminconducting-layer conductivity is not monotonic. The attenuation exceeds 20 dB/100 m around the semiconducting-layer conductivity equal to one millisiemen per meter and 1 000 siemens per meter. This slide represents the results of calculations for forty values of conductivity.

The figure on the right shows the dependence of the propagation velocity of a 30-megahertz signal on the semiconducting-layer conductivity. The vertical axis represents the signal propagation velocity in meters per microsecond. The horizontal axis represents the semiconducting-layer conductivity in siemens per meter. You can see the propagation velocity is 172 meters per microsecond when the semiconducting-layer conductivity is lower than about 10^{-4} (ten to the minus four) siemens per meter, drops to 135 meters per microsecond for conductivity between 10 millisiemens and 100 siemens per meter, and rises again to 170 meters per microsecond for conductivity above 10^4 (ten to the fourth) siemens per meter.

Let's move on to the next slide.

このスライドの左側の図をご覧になってください．電圧印加点から

2.1 発表用スライドの作成法と発表法

100 m 離れた点における周波数 30 MHz の信号の振幅の半導電層導電率への依存性を示しています．縦軸は，電圧振幅をボルト単位で示しています．横軸は，半導電層の導電率をジーメンス毎メートルで示しています．ご覧のように，信号の減衰の半導電層導電率への依存性は単調ではありません．半導電層の導電率が 1 mS/m 程度および 1 000 S/m 程度のときに，減衰は 20 dB/100 m を超えています．このスライドは，40 の導電率値に対する計算結果を示しています．

　右側の図は，周波数 30 MHz の信号の伝搬速度の半導電層導電率への依存性を示しています．縦軸は，電圧パルスの伝搬速度をメートル毎マイクロ秒単位で示しています．横軸は，半導電層の導電率をジーメンス毎メートルで示しています．ご覧のように，半導電層の導電率が 10^{-4} S/m 程度より低いときには，伝搬速度は 172 m/μs で，導電率が 10 mS/m から 100 S/m の範囲では，135 m/μs に落ち込み，導電率が 10^4 S/m 程度より高くなると，再び 170 m/μs に上昇しています．

　それでは，次のスライドに移りましょう．

〈補足〉 "Please take a look at …"「…をご覧になってください．」はよく使う表現です．

"Let's move on to …"「…に移りましょう．」は，スライドを進める際に有用な表現です．

「マイクロ秒」は，"micro-second" と記述されることはまれで，ほとんどの場合，"microsecond" と記述されます．"nanosecond"，"megahertz" などについても同様です．

2.1.7　考察の説明　*How to Explain the Discussions*

次は，得られた結果の考察について簡潔に説明します．たとえば図 2.8〜図 2.10 のようなスライドをスクリーンに表示し，次のように説明を行います．

　Let me first discuss the reason for the significant signal attenuation and dispersion when the semiconducting-layer conductivity is around one millisiemen per meter. This figure shows FDTD-computed waveforms of a 30-megahertz signal at distances of 20, 40, 60, 80, and 100 meters from the excitation point along the power cable with this semiconducting layer. You can see the signal suffers attenuation and dispersion with propagation. I think the attenuation and dispersion are

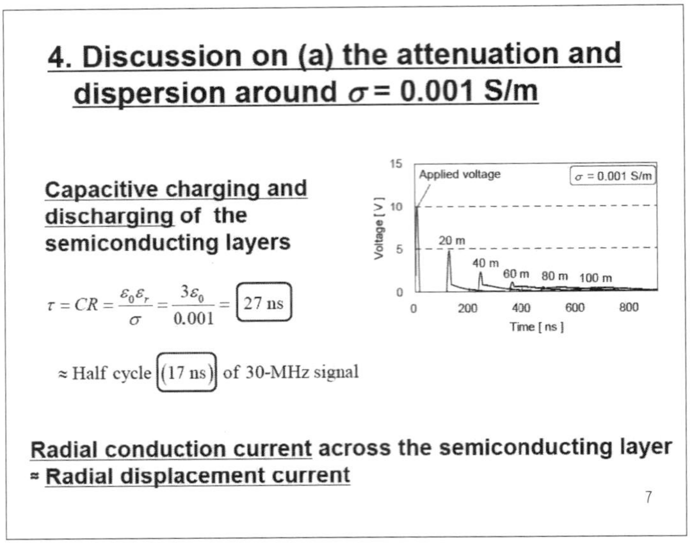

第2章　英語論文口頭発表法

> ### 4. Discussion on (a) the attenuation and dispersion around $\sigma = 0.001$ S/m
>
> **Capacitive charging and discharging of the semiconducting layers**
>
> $\tau = CR = \dfrac{\varepsilon_0 \varepsilon_r}{\sigma} = \dfrac{3\varepsilon_0}{0.001} = \boxed{27 \text{ ns}}$
>
> ≈ Half cycle $\boxed{(17 \text{ ns})}$ of 30-MHz signal
>
> **Radial conduction current** across the semiconducting layer
> ≈ **Radial displacement current**

図 2.8　検討結果を示したスライド（その 1）

due to capacitive charging and discharging in a radial direction of the semiconducting layers. This is because the time constant of the semiconducting layer of one millisiemen per meter is 27 nanoseconds. This is close to the half cycle of a 30-megahertz signal. In this condition, the magnitude of the radial conduction current across the semiconducting layer is close to that of the radial displacement current. In other words, the conductance of the semiconducting layer is close to its susceptance.

　まずは，半導電層の導電率が 1 mS/m の場合に生じる信号の著しい減衰と変わいの原因について検討を行います．こちらの図は，この半導電層をもつ電力ケーブルの電圧印加点から 20，40，60，80 および 100 m 離れた点における 30 MHz の信号の FDTD 計算波形を示しています．ご覧のように，信号は伝搬に伴い減衰および変わいしています．この減衰と変わいは半導電層の半径方向における充放電に起因したものであると考えられます．なぜなら，導電率 1 mS/m の半導電層の時定数は 27 ナノ秒であり，30 MHz の信号の半サイクルに近いからです．このとき，半径方向に流れる伝導電流の大きさと変位電流の大きさは同程度になります．このことは，半導電層のコンダクタンスとサセプタンスが同程度になるといい換えることもできます．

〈補足〉「先ほど示した図」などといわずに，考察に必要な図や重要な結果

は繰り返し示しましょう．

> Next, let me discuss the reason for the significant attenuation and dispersion when the semiconducting-layer conductivity is around 1 000 siemens per meter. This figure shows FDTD-computed waveforms of a 30-megahertz signal at distances of 20, 40, 60, 80, and 100 meters from the excitation point along the power cable with inner and outer semiconducting-layers of conductivity 1 000 siemens per meter. You can see the signal suffers similar attenuation and more severe dispersion with propagation than the layers with one millisiemen per meter conductivity. I think in this case the attenuation and dispersion are due to axial current propagation in the semiconducting layers, not radial charging and discharging. This is because the penetration depth for a semiconducting layer of conductivity 1 000 siemens per meter and frequency 30 megahertz is 2 millimeters, a value that is relatively close to the semiconducting-layer thickness.
>
> 次に，半導電層の導電率が 1 000 S/m の場合に生じる著しい減衰と変わいの原因について検討を行います．こちらの図は，内側および外側の半導電層の導電率が 1 000 S/m の電力ケーブルの電圧印加点から 20，40，60，80 および 100 m 離れた点における 30 MHz の信号の FDTD 計算波形を示しています．ご覧のように，伝搬に伴い，1 mS/m の場合と

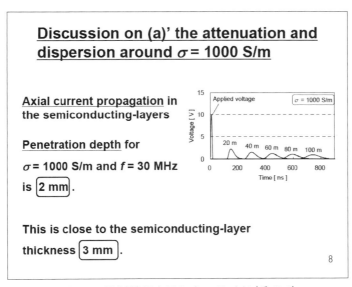

図 2.9　検討結果を示したスライド（その 2）

同じように信号は減衰し，1 mS/m の場合よりも大きく変わいしています．この場合の減衰と変わいは，半径方向の充放電ではなく，軸方向電流が半導電層を流れることに起因したものであると考えられます．なぜなら，導電率 1 000 S/m，周波数 30 MHz での透過深さが，半導電層の厚みに比較的近い 2 mm であるからです．

Next, I would like to analyze the propagation velocity in more detail. This figure again shows the dependence of the propagation velocity of a 30-megaherz signal on the semiconducting-layer conductivity, varying from 172 meters per microsecond at low conductivity to 135 meters per microsecond over a broad range of three orders of magnitude around one siemen per meter, and then increasing to 170 meters per microsecond for high conductivity.

When the semiconducting-layer conductivity is lower than about 10^{-4} siemens per meter, the semiconducting layers behave as insulators. When the semiconducting-layer conductivity is higher than about 10^4 siemens per meter, the semiconducting layers behave as conductors. Therefore, in these two cases, the signal propagation velocity is theoretically determined by the permittivity values of the insulating and semiconducting layers or that of the insulating layer, respectively. The value is 300 meters per microsecond divided by the square root of

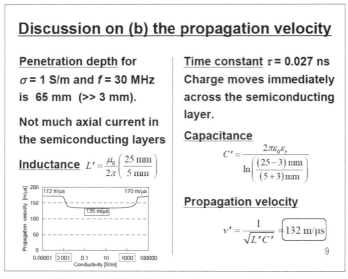

図 2.10　検討結果を示したスライド（その 3）

relative permittivity, 3, which gives 173 meters per microsecond. This estimate agrees well with the FDTD-computed velocity.

When the semiconducting-layer conductivity is one siemen per meter, the penetration depth at 30 megahertz is 65 millimeters. This is much larger than the 3-millimeter thickness of each semiconducting layer. Therefore, the axial current propagates mainly on the surface of the core conductor and the inner surface of the sheath conductor, and the magnetic field energy is stored between the surface of the core conductor and the inner surface of the sheath conductor. The corresponding inductance per unit length is given by this expression for coaxial geometry. On the other hand, since the time constant of the one-siemen-per-meter semiconducting layer is 0.027 nanosecond, charge moves immediately across the semiconducting layer. Therefore, electric field energy is stored in the insulating layer, and the corresponding capacitance per unit length is given by this standard expression, also for coaxial geometry. From these inductance and capacitance, we can theoretically get a propagation velocity of 132 meters per microsecond. The FDTD-computed velocity agrees well with this theoretical value.

次に，伝搬速度について，さらに詳しく解析します．こちらの図は，再度，30 MHz 信号の伝搬速度の半導電層導電率への依存性を示しています．低い導電率において 172 m/μs であったのが，1 S/m 付近の3桁を超える広い範囲において 135 m/μs になり，それから高い導電率において 170 m/μs に上昇しています．

半導電層の導電率が 10^{-4} S/m 程度以下の場合には，半導電層は絶縁体として振る舞います．半導電層の導電率が 10^4 S/m 程度以上の場合には，半導電層は導体として振る舞います．したがって，これらの場合には，理論的には，伝搬速度は絶縁層と半導電層の比誘電率あるいは絶縁層の比誘電率で決まります．その値は，300 m/μs 割る比誘電率3の平方根，つまり 173 m/μs となります．この推定値は FDTD 計算で得た伝搬速度に良好に一致します．

半導電層の導電率が 1 S/m 程度の場合，30 MHz における透過深さは 65 mm となります．この値は，半導電層の厚み 3 mm に比べて十分大きいといえます．したがって，軸方向電流は主として中心導体の表面とシース導体の内側表面を流れ，磁界エネルギーは中心導体とシース導体

間に蓄えられます．これに対応する単位長あたりのインダクタンスは，同軸形状に対して，この式で与えられます．一方，導電率が 1 S/m 程度の半導電層の時定数は 0.027 ns ですので，電荷は半導電層を即座に横切ります．したがって，電界エネルギーは絶縁層に蓄えられ，対応する単位長あたりのキャパシタンスは，この場合もまた同軸形状に対して，この標準的な式で与えられます．これらのインダクタンスとキャパシタンスから，伝搬速度 132 m/μs が理論的に得られます．FDTD 計算結果は，この理論値にほぼ一致しています．

2.1.8 まとめの説明
How to Explain the Summary or Conclusions

最後に，論文内容のまとめについて簡潔に説明します．たとえば図 2.11 のようなスライドをスクリーンに表示し，次のように説明を行います．

Finally, I would like to summarize my presentation. In this paper, we have studied the propagation characteristics of a 30-megahertz PLC signal along a power cable having two, 3-millimeter thick semiconducting layers using the FDTD method. We found a 30-megahertz PLC signal suffers significant attenuation and dispersion when the semiconducting-layer conductivity is around either one millisiemen per

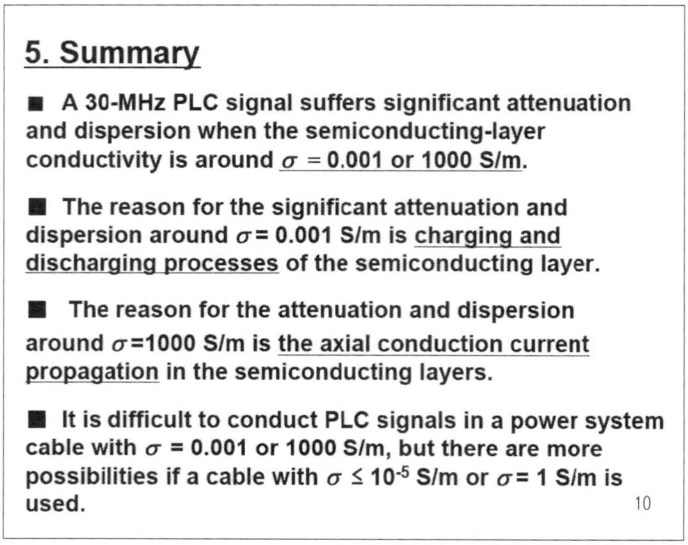

5. Summary

■ A 30-MHz PLC signal suffers significant attenuation and dispersion when the semiconducting-layer conductivity is around σ = 0.001 or 1000 S/m.

■ The reason for the significant attenuation and dispersion around σ = 0.001 S/m is charging and discharging processes of the semiconducting layer.

■ The reason for the attenuation and dispersion around σ = 1000 S/m is the axial conduction current propagation in the semiconducting layers.

■ It is difficult to conduct PLC signals in a power system cable with σ = 0.001 or 1000 S/m, but there are more possibilities if a cable with $\sigma \leq 10^{-5}$ S/m or σ = 1 S/m is used.

図 2.11　論文内容のまとめを示したスライド

meter or 1 000 siemens per meter. The reasons for the significant signal attenuation and dispersion around one millisiemen per meter are charging and discharging processes in a radial direction of the semiconducting layers. The reason for the significant signal attenuation and dispersion around 1 000 siemens per meter is the axial conduction current propagation in the semiconducting layers. We conclude that it will be difficult to conduct PLC signals in a power system cable with semiconducting layers of conductivity one millisiemen per meter or 1 000 siemens per meter, but there are more possibilities if a semiconducting-layer conductivity lower than ten to the minus five siemens per meter or of about one siemen per meter is used.

Thank you very much for your kind attention.

最後に本発表をまとめたいと思います．本論文では，FDTD法を用いて，厚み3 mmの半導電層を2層有する電力ケーブル上での周波数30 MHzのPLC信号伝搬特性について検討を行いました．その結果，周波数30 MHzのPLC信号は，半導電層の導電率が1 mS/mまたは1 000 S/m程度のときに，著しい減衰と変わいを生じることがわかりました．導電率が0.001 S/mの場合に著しい減衰と変わいが生じるのは，半導電層の半径方向における充放電現象が原因です．導電率が1 000 S/mの場合に著しい減衰と変わいが生じるのは，半導電層に軸方向電流が流れることが原因です．以上より，1 mS/mまたは1 000 S/mの半導電層を有する電力ケーブルでは，PLC信号の伝導は困難となりますが，半導電層の導電率が10^{-5} S/mより低い場合あるいは1 S/m程度の場合には，PLC信号伝導の可能性はより高まるという結論になります．

ご清聴ありがとうございました．

〈**補足**〉 "Thank you very much for your kind attention."「ご清聴ありがとうございました．」はよく使われる表現です．

研究内容や成果を決められた時間内で的確に伝えられるように，リハーサルを何度も行いましょう．

第 2 章　英語論文口頭発表法

2.2　グラフや表の説明法
How to Explain Graphs and Tables

本節では，グラフや表の説明時に用いられる表現を紹介します．

2.2.1　グラフまたは表の表示
How to Show a Graph or a Table

スライド上のグラフ（または表）は，次のようにいいながら見せます．

> Please take a look at this graph（または table）.
>
> こちらの図（または表）をご覧ください．

〈補足〉 "Let's take a look at this graph." という場合もあります．
　線グラフは "line graph"，棒グラフは "bar graph"，円グラフは "pie chart（または graph）" といいます．

2.2.2　グラフの説明　*How to Explain a Graph*

たとえば，図 2.12 のグラフに対しては，対応する箇所をレーザポインタで示しながら，次のように説明します．

> 　This graph shows waveforms of a voltage pulse at distances of 20, 40, 60, 80, and 100 meters from the excitation point along a power cable. The applied voltage has a positive half cycle of a sine wave of frequency 30 megahertz, and its magnitude is 10 volts. We computed these waveforms using the FDTD method for a power cable with inner and outer semiconducting layers of conductivity 1 000 (one thousand) siemens per meter. The vertical axis represents the voltage in volts. The horizontal axis represents time in nanoseconds. As you can see, the magnitude of the voltage pulse decreases with increasing propagation distance.

2.2 グラフや表の説明法

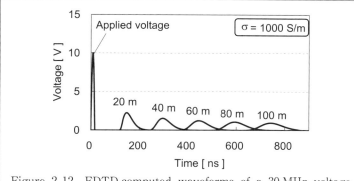

Figure 2.12 FDTD-computed waveforms of a 30-MHz voltage pulse at different distances from the excitation point along a power cable with inner and outer semiconducting layers of conductivity 1 000 S/m.

図 2.12　グラフの例

　このグラフは，電圧印加点から電力ケーブルに沿って 20，40，60，80 および 100 m 離れた点における電圧パルス波形を示しています．印加電圧は周波数 30 MHz の正弦波の正半波で，その振幅は 10 V です．これらは，電力ケーブルの内側および外側半導電層の導電率が 1 000 S/m の場合の FDTD 法による計算結果です．縦軸は電圧をボルト単位で示しています．横軸は時間をナノ秒単位で示しています．ご覧のように，電圧パルスの振幅は伝搬に伴って減衰しています．

〈補足〉 "This graph shows...", "The vertical axis represents... The horizontal axis represents...", "As you can see,..." はよく用いられる表現です．

　図の横軸，縦軸の "Time [ns]"，"Voltage [V]" を "label" といいます．

　グラフの太い実線（ ——— ）は "thick solid line"，細い実線（ ――― ）は "thin solid line"，点線（ ……… ）は "dotted line"，破線（ ‑‑‑‑‑‑ ）は "broken line"，長破線（ － － － － ）は "dashed line"，一点鎖線（ ‑・‑・‑ ）は "dashed-dotted line" といいます．たとえば，「この破線は...です．」は "The broken line indicates（または shows）..." といいます．

　記号については，" * " は "asterisk"，"・" は "dot"，"○" は "circle"，"◇" は "diamond"，"□" は "square"，"△" は "triangle"，"●" は "solid circle" といいます．たとえば，「●は...です．」は "The solid circle denotes..." といいます．

2.2.3 表の説明　*How to Explain a Table*

たとえば，表2.2に対しては，対応する箇所をレーザポインタで示しながら，次のように説明します．

> This table shows characteristic-impedance values of an infinitely long horizontal conductor that we calculated for different sets of conductor radius and height. The leftmost column indicates the conductor radius. We consider 10 and 20 millimeters for the radius. The middle column indicates the height. We consider 1 and 2 meters. The rightmost column indicates the characteristic impedance calculated. As we can see, the characteristic impedance ranges from 280 to 360 ohms.
>
> この表は，異なる半径および高さの無限長水平導体の特性インピーダンスの計算結果を示しています．（最も）左側の列は導体半径を示しています．今回は，10および20 mmとしています．中央の列は導体高さを示しています．今回は1および2 mとしています．（最も）右側の列は特性インピーダンスの計算結果を示しています．ご覧のように，特性インピーダンスは280～360 Ωの範囲にあります．

〈補足〉　表の行（横）を"row"，表の列（縦）を"column"といいます．最も左側，最も右側は"leftmost"，"rightmost"といいます．"range from A to B"（AからBの範囲にある）はよく使う表現です．

表よりもグラフのほうが短時間で情報を伝えられるため，発表用スライドにおいては，グラフ化できるものはグラフにして示すほうがよいといわれています．

表2.2　表の例

Table 2.2　Characteristic impedance of an infinitely long horizontal conductor calculated for different sets of conductor radius and height

Radius, mm	Height, m	Characteristic impedance, Ω
10	1.0	320
20	1.0	280
10	2.0	360
20	2.0	320

2.3 数と数式の読み方
How to Read Numbers and Mathematical Expressions

数(numbers)や数式(expressions)は，日本語論文におけるのとほぼ同様に英語論文においても表記されます．このため，英語論文を執筆する段階では，それらの読み方に留意する必要はないかもしれません．しかし，数や数式を含む論文を英語で口頭発表する場合には，それらを声に出して読まなければなりません．本節では，数と数式の英語での読み方を説明します．

2.3.1 基数と序数　*Cardinal Numbers and Ordinal Numbers*

> 12 300 456
>
> Twelve million, three hundred thousand, four hundred fifty-six

〈補足〉　基数を読む際には，1の位から3桁ずつ区切ります．このとき，各区切りのところで，"thousand"（千），"million"（百万），"billion"（十億）などの単位をつけます．これらは複数形にはしません．余談ですが，英語圏の国々では，数字の3桁ごとの区切りにカンマが用いられていますが（例 12,300,456），ドイツなどの非英語圏の国々では，小数点にカンマが用いられるため，3桁の区切りにはピリオドかスペースが用いられています．

長さ，質量，時間などの物理量が対象の場合には，千倍，百万倍，十億倍を表す接頭語 "kilo"，"mega"，"giga" を物理単位につけてコンパクトに表現するのが一般的です．たとえば，5万6千メートルは 56 km と略記し，"fifty-six kilometers" と読みます．2語から構成される99以下の数は，この例からもわかるように，"fifty-six" のように，ハイフンでつないで書かれます．なお，"giga" は「ギガ」と発音されるのが一般的ですが，まれに「ジガ」と発音される場合があります（フランス語やイタリア語では「ジガ」と発音）．整数は "whole number" または "integer"，素数は "prime number" といいます．

> 123rd

> One hundred twenty-third

〈補足〉 100th は "hundredth", 1 000th は "thousandth" と読みます.

2.3.2 小数と分数 *Decimals and Fractions*

> 123.456
>
> One hundred twenty-three point four five six

〈補足〉 小数点までの整数部分 (whole number part) を基数として読み, 小数点を "point" と読みます. 小数部分 (fractional part) は 1 桁ずつ順番に読みます. 循環小数 (recurring decimals) を読む場合には, 「循環する」を意味する形容詞 "recurring" を後ろにつけます. たとえば, $0.\dot{3}$ は "zero point three recurring" と読み, $2.\dot{4}2\dot{3}$ は "two point four hundred twenty-three recurring" と読みます.

「123.456 を小数第 1 位で四捨五入しなさい.」という場合には, "Round off 123.456 to unit." となります. 「123.456 を小数第 2 位で四捨五入しなさい.」という場合には, "Round off 123.456 to the nearest tenth." または "Round off 123.456 to one decimal place." となります. 「123.456 を有効桁数が 4 になるように四捨五入しなさい.」という場合には, "Round off 123.456 to four significant figures." となります. 「123 456 を 100 の位で四捨五入しなさい.」という場合には, "Round off 123 456 to the nearest thousand." となります.

一般に, 1, 0.1, 0.01 などは単数扱いされ, それら以外は複数扱いされます. たとえば, 摂氏マイナス 0.01 度 (−0.01 ℃) は "minus zero point zero one degree Celsius" といいます. 一方, 摂氏 0 度 (0 ℃) は "zero degrees Celsius", 摂氏マイナス 2 度 (−2 ℃) は "minus two degrees Celsius", 華氏 32 度 (32 ℉) は "thirty-two degrees Fahrenheit", 絶対温度 273.15 度 (273.15 K) は "two hundred seventy-three point one five kelvins" といいます.

> $\dfrac{24}{5}$
>
> Twenty-four fifths   または   Twenty-four over five

〈補足〉 分数 (fraction) を読む場合には, 分子 (numerator) を基数詞, 分

母(denominator)を序数詞で読みます．分子が2以上の場合には分母の序数を複数形にします．たとえば，24/5 は，1/5(one fifth) が 24 あると考え，"Twenty-four fifths" とします．分子と分母の間に "over" を入れて，分母も基数として読む場合もあります．

$4\frac{4}{5}$ と帯分数(mixed fraction)で表されている場合には，"four and four fifths" と読みます．

有理数は "rational number"，無理数は "irrational number" といいます．

「5 の逆数は 1/5 です．」という場合には，"The reciprocal of five is one fifth." となります．

2.3.3 n 乗根，累乗，対数
nth Root, Powers, and Logarithms

$\sqrt[n]{x}$

The nth root of x

〈補足〉 $\sqrt{x}$ は "the square root of x" と読み，$\sqrt[3]{x}$ は "the cube root of x" と読みます．このように，論文中において，変数は斜体で表記します．

---
x^n

x to the nth　または　x to the nth power

〈補足〉 x^{-n} は "x to the minus n" と読みます．x^2 は "x squared" または "the square of x" と読みます．x^3 は "x cubed" と読みます．

e^x は "e to the x" と読みます．$\exp(x)$ は "the exponential of x" と読みます．

「x を 4 乗しなさい．」という場合には，"Raise x to the fourth power." となります．

$\ln x$

The natural log of x　または　ln x（ロン エクス）

〈補足〉 $\log_{10} x$ は "the common log of x" または "the log base ten of x" と読みます．論文中において，関数名は直立体で表記します(例　ln, exp, sin など)．

2.3.4　足し算と引き算　*Addition and Subtraction*

$$x+\sqrt{x+y}=z$$

x plus the square root of the sum of x and y equals z.
または　x plus the square root of x plus y equals z.

〈補足〉 "z" は「ズィー」と発音されます．ただし，イギリス英語では「ゼッド」と発音されます．

5+2=7 を「5 に 2 を足しなさい，そうすれば 7 が得られます．」という場合には，"Add two to five, and you get seven." となります．

$$x_1-(-x_2)=\sqrt{11-2}$$

x sub one minus minus（または negative）x sub two equals the square root of the quantity of eleven minus two.

〈補足〉 上式に "−" は三つあります．1 番目と 3 番目の "minus" は前置詞で，2 番目の "minus" は形容詞です．したがって，2 番目は "negative" と読むこともできます．"the quantity of" が抜けると，上式の右辺は $\sqrt{11}-2$ と誤解されてしまう可能性がありますが，スライドなどで上式を見せながら読む場合には，"the quantity of" を省略しても誤解される可能性は低いでしょう．

5−2=3 を「5 から 2 を引きなさい，そうすれば 3 が得られます．」という場合には，"Subtract two from five, and you get three." となります．

x_1 は "x sub one" と読みます．"sub" は下付き文字 "subscript" を示しています．累乗を意味しない上付き文字(superscript)については，たとえば，x^k は "x super k" と読みます．

x' は "x prime"（"x dash" と読む場合もありますが，"dash" は「―」も意味します），x'' は "x double prime"，$\bar{x}$ は "x bar"，$\hat{x}$ は "x hat"，$\tilde{x}$ は "x tilde" と読みます．

2.3.5 掛け算と割り算　*Multiplication and Division*

$x \times y = z$

x times y equals z.　または　x multiplied by y equals z.

〈補足〉　$xy=z$ も同様に読みますが，"xy equals z." と読むこともあります．$3x^2+13x+4=0$ は，"Three times x squared plus thirteen times x plus four equals zero" と読むこともありますし，"times" を省略して読むこともあります．

3×10^4 は "three times ten to the fourth" と読みます．

$5 \times 2=10$ を「5 に 2 を掛けなさい，そうすれば 10 が得られます．」という場合には，"Multiply five by two, and you get ten." と読みます．

$x^2 - y^2 = (x+y)(x-y)$

x squared minus y squared equals the sum of x and y times the quantity of x minus y.

〈補足〉　電話などで，括弧の位置を相手に，明確に伝えなければならない場合には，"x squared minus y squared equals parenthesis open x plus y parenthesis close times parenthesis open x minus y parenthesis close." と読みます．括弧 "(" は "parenthesis open" または "initial parenthesis"，")" は "parenthesis close" または "final parenthesis" と読みます．"( )" は複数形にして，"parentheses" と読みます．

なお，"[ ]" は "square brackets"，"{ }" は "braces" と読みます．

$\cosh^2\theta - \sinh^2\theta = 1$

The hyperbolic cosine squared of theta minus the hyperbolic sine squared of theta equals one.

または　cos h squared theta minus sine h squared theta equals one.

〈補足〉　θ はギリシャ文字(Greek letter)で，シータと読みます．**表2.3**にギリシャ文字の英語綴りと読み方をまとめておきます．

表2.3　ギリシャ文字の英語綴りと読み方

大文字	小文字	英語綴り	読み方	大文字	小文字	英語綴り	読み方
A	α	alpha	アルファ	N	ν	nu	ニュー
B	β	beta	ベイタ	Ξ	ξ	xi	クシー
Γ	γ	gamma	ガマ	O	o	omicron	オミクロン
Δ	δ	delta	デルタ	Π	π	pi	パイ
E	ε	epsilon	エプシロン	P	ρ	rho	ロー
Z	ζ	zeta	ジータ	Σ	σ	sigma	シグマ
H	η	eta	イータ	T	τ	tau	タウ
Θ	θ	theta	シータ	Y	υ	upsilon	ユープシロン
I	ι	iota	アイオタ	Φ	ϕ	phi	ファイ
K	κ	kappa	カッパ	X	χ	chi	カイ
Λ	λ	lambda	ラムダ	Ψ	ψ	psi	プサイ
M	μ	mu	ミュー	Ω	ω	omega	オメガ

$$x \div y = z$$

x divided by y equals z.

〈補足〉　$a/b=c$ も同様に読む場合もありますし，"x over y equals z" と読む場合もあります．$12 \div 5 = 2 \text{ r } 2$ (商2 余り2)は "Twelve divided by five is two with a remainder of two" と読みます．商は "quotient" といいます．

$12 \div 5 = 2.4$ を「12を5で割りなさい，そうすれば2.4が得られます．」という場合には，"Divide twelve by five, and you get two point four." と読みます．

2.3.6　微分と積分　*Differential Calculus and Integral Calculus*

$$\frac{\mathrm{d}f(t)}{\mathrm{d}t} = 2t^2 + 3$$

The derivative of f of t with respect to t equals two times t squared plus three.

または　$\mathrm{d}f\,\mathrm{d}t$（ディエフ ディティ）equals two t squared plus three.

〈補足〉 2階微分の場合には "the second derivative", n 階微分の場合には "the nth derivative" と読みます.

$f''(t) = \dfrac{\mathrm{d}^2 f(t)}{\mathrm{d}t^2}$ は "f double prime of t equals d (ディ) squared f of t d (ディ) t squared." と読みます.

Δf は "delta f" または "finite difference of f" と読みます.

$$\Delta = \frac{\partial^2}{\partial x^2} + \frac{\partial^2}{\partial y^2} + \frac{\partial^2}{\partial z^2}$$

The Laplacian equals the second partial with respect to x plus the second partial with respect to y plus the second partial with respect to z.

〈補足〉 $\dfrac{\partial^2}{\partial x^2}$ は "∂ (ダイ) squared ∂ (ダイ) x squared" とも読まれます.

$$\int_{t_1}^{t_2} f(t)\,\mathrm{d}t = T$$

The definite integral from t sub one to t sub two of f of t with respect to t equals capital T.

〈補足〉 大文字と小文字を区別する必要がある場合には, 大文字に "upper case" または "capital" を, 小文字に "lower case" または "small" をつけて読みます.

$$\int x \sin^{-1} x\,\mathrm{d}x$$

The indefinite integral of x times inverse sine of x with respect to x.

〈補足〉 $\int x \arcsin x\,\mathrm{d}x$ の場合には, "the indefinite integral of x times arc sine of x with respect to x" と読みます.

$\iint$ は "double integral" (2重積分) と読み, $\oint$ は "circuital integral" (周回積分) と読みます.

2.3.7 数列，級数，極限　*Sequences, Series, and Limits*

$$a_n = ar^{n-1}$$

a sub *n* equals *a* times *r* to the *n* minus one.

〈補足〉　等比数列は "geometric sequences"，等差数列は "arithmetic sequences" といいます．数学的帰納法は "mathematical induction" といいます．

$$\sum_{k=1}^{n} a_k \equiv a_1 + a_2 + a_3 + \ldots + a_n$$

The sum（または summation）of *a* sub *k* from *k* equals one to *n* is defined to be *a* sub one plus *a* sub two plus *a* sub three plus, and so on, plus *a* sub *n*.

$$n! = \prod_{k=1}^{n} k$$

n factorial equals the product from *k* equals one to *n* of *k*.

〈補足〉　階乗は "factorial" といいます．

$$\lim_{n \to \infty} a_n = 10$$

The limit of sequence *a* sub *n* as *n* goes to（または approaches）infinity is ten.

2.3.8 順列，組合せ，確率
Permutations, Combinations, and Probabilities

$$_nP_r = \frac{n!}{(n-r)!}$$

> The number of permutations of r from n equals n factorial over n minus r factorial.

$$_nC_r = \frac{n!}{r!(n-r)!}$$

> The number of combinations of r from n equals n factorial over r factorial times n minus r factorial.

$$P(A \cup B) = P(A) + P(B) - P(A \cap B)$$

> The probability of the union of A and B equals the probability of A plus the probability of B minus the probability of the intersection of A and B.

〈補足〉 "∪" は「和集合」を "∩" は「積集合」を表す記号です．

$$p(x) = \frac{1}{\sqrt{2\pi}\sigma} e^{-\frac{(x-\mu)^2}{2\sigma^2}}$$

> p of x equals one over the square root of two pi times sigma times e to the minus square of x minus mu divided by two sigma squared.

〈補足〉 "sigma" と "times" の間および "square of" と "x" の間で一呼吸おきます．σ は標準偏差，μ は期待値です．

2.3.9 不等式，比較　*Inequalities and Comparison*

> $x + y \leq z$
>
> x plus y is less（または smaller）than or equal to z.

〈補足〉 $x+y \geq z$ は "x plus y is greater than or equal to z" と読み，$x+y \gg z$ は "x plus y is much greater than z" と読みます．また，$x \approx z$ は "x is nearly（または approximately）equal to z"，$x \sim z$ は "x is asymptotic to z"

と読みます．$x \propto z$ は "x is directly proportional to z" または "x is in direct proportion to z" と読みます．$x \propto 1/z$ は "x is inversely proportional to z" または "x is in inverse proportion to z" と読みます．$x \equiv y$ は "x is identical to y" と読みます．

∵ $x=z$ は "Since x equals z"，∴ $x=z$ は "Therefore, x equals z" と読みます．

$x : y$

The ratio of x to y.

〈補足〉 $3:6=1:2$ は "Three is to six as one is to two" と読みます．

2.3.10 複素数　*Complex Numbers*

$$\mathrm{Re}(z) = \frac{z + \bar{z}}{2}$$

The real part of (a complex number) z equals z plus z bar（または its conjugate）over two.

$$\sin(z) = \frac{e^{jz} - e^{-jz}}{2j}$$

Sine of z equals e to the j z minus e to the minus j z over 2 j.

〈補足〉 "j" と "z"，"2" と "j" の間に "times" を入れてもよいですが，省略して読まれることが多いようです．三角関数は "trigonometric function" といいます．

$$\cos\theta + j\sin\theta = e^{j\theta}$$

Cosine of theta plus j times sine of theta equals e to the j theta.

〈補足〉 "Euler's formula"（オイラーの式）のように，人名の語尾にアポス

トロフィエスをつけ所有格で使う場合には，固有名詞扱いとなり冠詞をつけません．アポストロフィエスをつけずに定冠詞 "the" をつけて，"the Euler formula" という場合もありますが，"Euler's formula" がより頻繁に用いられています．この場合，"the" は普通名詞の "formula" にかかっていると考えます．"the Euler's formula" とはいいません．

2.3.11 ベクトル *Vectors*

$$\nabla \times \boldsymbol{E} = -\frac{\partial \boldsymbol{B}}{\partial t}$$

The curl of vector E equals minus dye B dye t.　または　Del cross E equals minus dye B dye t.

〈補足〉 $\nabla \times \boldsymbol{E}$ は "The rotation of E" と読む場合もあります．なお，$A \times B$ は "The vector（または cross, outer）product of vectors A and B" と読みます．"dB/dt" は「ディビー ディティ」と読みます．

$$\nabla \cdot \boldsymbol{B} = 0$$

The divergence of vector B equals zero.　または　Del dot B equals zero.

〈補足〉 $\boldsymbol{A} \cdot \boldsymbol{B} = |A||B| \cos \theta$ は "The scalar（または dot, inner）product of vectors A and B equals the norm of A times the norm of B times cosine of theta" と読みます．

$$\boldsymbol{E} = -\nabla V - \frac{\partial \boldsymbol{A}}{\partial t}$$

Vector E equals minus grad V minus dye A dye t.

$$\int_S \boldsymbol{D} \cdot \boldsymbol{n} \, \mathrm{d}S = Q$$

The surface integral of the scalar product of D dot n over S equals Q.

2.3.12 行列，行列式　*Matrices and Determinants*

$$(a_1, a_2, a_3)\begin{pmatrix} b_1 \\ b_2 \\ b_3 \end{pmatrix} = a_1 b_1 + a_2 b_2 + a_3 b_3$$

The one by three matrix *a* sub one, *a* sub two, *a* sub three times the three by one matrix *b* sub one, *b* sub two, *b* sub three equals *a* sub one times *b* sub one plus *a* sub two times *b* sub two plus *a* sub three times *b* sub three.

〈補足〉「1行3列の行列」を "one by three matrix" といいます．「行列 A の2行1列目の要素は a_{21} です．」は，"The element of matrix A in the second row and first column is *a* sub two one." といいます．

$$\begin{pmatrix} a & b \\ c & d \end{pmatrix}^{-1} = \frac{1}{\begin{vmatrix} a & b \\ c & d \end{vmatrix}} \begin{pmatrix} d & -b \\ -c & a \end{pmatrix}$$

The inverse of the two by two matrix *a*, *b*, *c*, *d* equals one over its determinant times the two by two matrix *d*, minus *b*, minus *c*, *a*.

〈補足〉A^{-1} は "the inverse of matrix A"，A^T は "the transpose of matrix A"，det A は "the determinant of matrix A" と読みます．

2.4　質問への対処法
How to Handle Questions

本節では，質問への対処時に用いられる表現を示します．質問は注意深く聞き，丁寧に答えましょう．

2.4.1　質問への応答法　*How to Answer Questions*

口頭発表が終了すると，座長は次のようにいって，質疑応答を開始します．

2.4 質問への対処法

> Thank you for your interesting presentation. This paper is now open for discussion. Does anyone have any questions or comments?
>
> 興味深いご発表をありがとうございました．それでは，質疑応答の時間に入ります．ただ今のご発表に対しまして，ご質問またはコメントはございませんでしょうか．

質問は，次のような質問者の自己紹介から始まります．

> I'm John Smith from the University of Southeastern America. I have enjoyed your presentation. Please let me ask one question. Your conclusion is based on the results calculated for a single-phase cable. I wonder if your conclusion could be applied to a three-phase cable.
>
> 南東アメリカ大学のジョン・スミスです．ご発表を興味深く聞かせていただきました．一つ質問があります．あなたの結論は単相ケーブルを対象とした計算結果に基づいています．この結論を三相ケーブルにも適用できますでしょうか．

この質問に対しては，たとえば，次のように答えます．

> Thank you. Attenuation and dispersion of voltage pulse don't depend on the number of phases, only on the conductivity and thickness of semiconducting layers. Therefore, I think we could apply our conclusion to a three-phase cable.
>
> ご質問ありがとうございます．電圧パルスの減衰と変わいは相数には依存せず，半導電層の導電率と厚さに依存します．したがって，私たちの結論は三相ケーブルにも適用できると思います．

別の質問例を示します．

> I'm John Smith from the University of Southeastern America. Thank you for your great presentation. I have one question. Could you please let me know the specification of your PC you used in this work?

第2章　英語論文口頭発表法

> 　南東アメリカ大学のジョン・スミスです．素晴らしいご発表をありがとうございました．一つ質問があります．今回の研究で使用したパソコンの仕様を教えていただけませんでしょうか．

この質問に対しては，たとえば，次のように答えます．

> 　Thank you. My PC has a three-gigahertz dual-core processor, a two-megabyte cache memory, and a three-gigabyte RAM (random-access memory).
>
> 　ご質問ありがとうございます．使用しましたパソコンは，3 GHz のデュアルコアプロセッサ，2 MB のキャッシュメモリ，3 GB の RAM を有しております．

〈補足〉「この PC を用いた場合の計算時間は，1 ケースで約 5 時間でした．」は，"The computation time for one case was about five hours when we used this PC." といいます．

別の質問例を示します．

> 　I'm John Smith from the University of Southeastern America. Thank you for your interesting presentation. I am skeptical about the accuracy of your computed results. Could you please tell us about the accuracy of your computations?
>
> 　南東アメリカ大学のジョン・スミスです．興味深いご発表をありがとうございました．あなたの計算結果の精度に疑問があります．計算精度についてお話いただけませんでしょうか．

この質問に対しては，たとえば，次のように答えます．

> 　Thank you. In answer to your question, we have compared our computed results with the corresponding measurements, and found excellent agreement. The maximum difference in peak voltages between the computed and measured voltages is about 5 %, which is near the limit of our measurement accuracy. Therefore, we think that

our computed results are sufficiently accurate. I hope I've answered your question.

　ご質問ありがとうございます．ご質問にお答えいたしますと，私たちは計算結果を対応する実測結果と比較し，良好に一致することを確認しております．計算結果と実測結果との電圧ピーク値の差が，最大で，測定精度の上限に近い約5％となっております．したがって，私たちの計算結果は十分な精度を有していると考えております．ご質問の答えになっていればよいのですが．

〈補足〉 "In answer to your question, …" は応答時によく使う表現です．この質問は短いですが，長い質問を受けた場合，質問内容を次のように確認してから回答すると，会場のほかの方にも質疑応答内容がわかりやすくなります．"My understanding of your question is about the accuracy of our computations of peak voltage. Am I correct in understanding your question?" あるいは "If I understood correctly, your question is about the accuracy of our computations of peak voltage." 質問に的確に答えられているか否かを質問者に確認する場合，"Did I answer your question?" ということもあります．

　計算精度の検証を今後行う予定である場合には，たとえば次のように答えます．

　Thank you. We are interested in that point too, and are carrying out experimental verifications using an actual power cable.

　ご質問ありがとうございます．その点については私たちも関心があり，実際の電力ケーブルを用いて実験的に検証を行う予定です．

質問が聞き取れなかった場合には，次のようにいいましょう．

　I couldn't catch your question well. Could you please repeat it?

　ご質問を十分に聞き取れませんでした．もう一度繰り返していただけませんでしょうか．

〈補足〉 "catch" の代りに "follow" を用いることもできます．特に，質問が長く複雑なため聞き取れなかった場合には，"follow" がより適しています．

2.4.2 コメントへのお礼の仕方　*How to Thank for Comments*

賞賛のコメントをいただいた場合には，次のようなお礼をいいましょう．

> Thank you very much for your compliment（または kind words）.
>
> お褒めいただき，ありがとうございます．

有用なご提案をいただいた場合には，次のようにいいましょう．

> Thank you very much for your useful suggestion. The FDTD method is ideal for looking into this aspect.
>
> 有用なご提案をありがとうございます．FDTD 法は，そのことについて検討を行うのに理想的です．

〈補足〉「その点では，私もあなたと同意見です．」は，"I agree with you on that point." または "I'm with you on that point." といいます．「その点では，私はあなたとは意見が異なります．」は，"I disagree with you on that point." といいます．

ご提案をいただいたことについてさらに詳しいお話を聞きたい場合には，次のようにいいましょう．

> Thank you very much for your intriguing（または interesting）suggestion. I would be happy if I could discuss it with you during the next coffee break.
>
> 興味深いご提案をありがとうございます．次の休憩時間にディスカッションすることができればと思っています．

2.4.3　むずかしい質問への対処法
How to Handle Difficult Questions to Answer

まだ結論の得られていないことに対する質問には，次のように答えましょう．

> Your question is very difficult for me to answer right now. It would require further research.
>
> ご質問は，今ここでお答えするのは非常にむずかしいと思います．（その答えを得るためには，）今後，さらなる研究が必要であると思います．

答えるのに十分な情報が手元にない場合には，次のようにいいましょう．

> I'm sorry, I don't have sufficient information (data) to answer your question right now.
>
> すみませんが，あなたのご質問にお答えするのに十分な情報（データ）が手元にございません．

どうしても答えられない場合には，次のようにいいましょう．

> I am sorry, I have no idea about it. Please see me afterwards so I can get the details.
>
> すみませんが，それについてはわかりません．後ほど私のところにきてください．そうすれば，私はご質問の詳細を知ることができます．

〈補足〉　言語（英語）の問題で，その場で答えられない場合には，"I'm sorry, I have a problem understanding spoken English." といいましょう．あるいは，"Let's get together with the translation program on my phone to make sure I understand your point". といいましょう．

2.5 国際会議での座長のことば
How to Talk as the Chairperson of a Session at an International Conference

本節では，国際会議の口頭発表セッションで一般的に述べられる座長のことばを説明します．セッションの流れや，座長が話すであろう内容がわかっていれば，聞き取らなければならない情報は，たとえば，講演時間，質疑応答時間，質問の内容だけになり，聞取りに使う神経も少なくてすみます．また，本節の内容は，将来，座長を任されたときにも，役立つと思います．

2.5.1 セッション前の準備 *How to Prepare for a Session*

座長は，担当セッションで発表される予定の論文を事前に読み，各論文に対して簡単な質問を準備しておく必要があります．セッションの最初の論文に対する最初の質問は，そのセッションでの以降の質疑応答の論調や雰囲気を左右する場合があります．したがって，会場からの質問がない場合には，座長は建設的で比較的簡単な（しかし，聞くまでもないような明白なものではない）質問をするように心掛けます．

座長は，セッション会場に少し早く入り，発表に使われるパソコンに各発表者がパワーポイントスライドのファイルをコピーし，適切に動作することを確認するように促します．また，発表用スライド操作用のリモコンやマイクロフォンの使い方，発表者が自身のパソコンを使用する場合には，プロジェクタとの接続法についても事前確認してもらいます．なお，会場の出席者はカジュアルな服装である場合が多いですが，座長は，セッション会場に入室してきた発表者や会場係（パソコン，プロジェクタ，マイクロフォンや照明の管理をされる方）にも容易に見分けがつくように，ビジネススーツを着ておくのが親切です．

2.5.2 セッション開始時のことば *How to Start a Session*

口頭発表セッションの開始にあたって，座長は次のような挨拶をします．

Good morning (good afternoon) ladies and gentlemen. Welcome to the session on Lightning Electromagnetics. I am Jane Doe from the

University of Southeastern America. I am going to chair this session. In this session, we have six paper presentations. Each speaker will have fifteen minutes for his or her presentation and five minutes for questions and answers. I am sorry but I will ask a speaker to stop the presentation if he or she exceeds this time limit. I will indicate twelve minutes and fifteen minutes by a bell.

Let's move on to the first paper. The title of the paper is an electromagnetic model of the lightning return stroke. The authors are Ms. Hanako Denki and Professor Taro Denshi of the University of Southern Kyoto, Japan. The paper will be presented by Ms. Hanako Denki.

皆様，おはようございます(こんにちは)．雷電磁気学に関する本セッションにようこそお越しくださいました．私は南東アメリカ大学のジェーン・ドゥです．このセッションの座長を務めさせていただきます．このセッションでは，6件の論文発表を予定しております．各講演者の持ち時間は，発表15分，質疑応答5分となっております．すみませんが，この制限時間を超えた場合には，発表を打ち切らせていただきます．発表開始後12分と15分に，ベルで時間をお知らせいたします．

それでは，最初の講演に移りましょう．最初の論文のタイトルは帰還雷撃の電磁界モデルです．著者は，南京都大学の電気花子さん，電子太郎教授です．講演は，電気花子さんにより行われます．

〈補足〉 各セッションに割り当てられた時間はかぎられていますので，上述のように，各発表の制限時間を明確に伝えておく必要があります．

2.5.3 発表終了時と質疑応答時のことば
How to Thank a Speaker and Invite Questions

各講演者の口頭発表が終了したら，座長は次のようにいって，質疑応答を開始します．

Thank you for your interesting presentation. This paper is now open for discussion. Does anyone have any questions or comments?

興味深いご講演をありがとうございました．それでは，質疑応答の時

間に入ります．ただ今の講演に対しまして，ご質問またはコメントはございませんでしょうか．

〈補足〉 "Are there any questions or comments?"，簡略化して "Any questions or comments?" という場合もあります．講演の途中で質問がでた場合には，次のようにいいます． "There will be time for questions. Please hold your questions until then."

最初の質疑応答の後，まだ質疑応答時間が残っている場合には，座長は次のようにいって，会場からの質問を促します．

Does anyone have any further questions?

ほかにご質問はございませんでしょうか．

講演者が質問内容を理解できていない場合，座長は次のようにいって，質問内容を別のことばで説明したり，質問をゆっくり繰り返したりします．

Please let me step in here, and allow me to reword the question for the speaker.

少し発言させていただきます．質問内容を別のことばで説明させてください．

同じ人が続けて質問しようとした場合には，座長は次のようにいって，挙手しているほかの人の質問に移ります．

Others have questions. I will get back to you if time allows.

ほかの方も質問がございます．時間が許せば，あなたに質問をお願いいたします．

発表が制限時間を超えた場合には，座長は講演者に次のようにいって，発表を打ち切ります．

I am afraid that time is almost up. Could you please finish your presentation?"

> すみませんが，もう時間がございません．発表を打ち切ってくださいませんでしょうか．

〈補足〉 "It is already twenty minutes. Please conclude your presentation?" という場合や "You have two more minutes. Please skip to the conclusions". という場合もあります．

2.5.4 次の講演に移るときのことば
How to Move on to the Next Presentation

講演が終わったら，次の講演に移ります．その際，座長は次のようにいいます．

> If there are no further questions, I would like to thank Ms. Denki for her nice presentation. Now, let's move on to the next presentation. The title of the paper is an engineering model of the lightning return stroke. The author and presenter is Professor Taro Denshi of the University of Southern Kyoto, Japan.
>
> ほかにご質問がなければ，素晴らしいご講演に対して電気さんに感謝したいと思います．それでは次の講演に移りましょう．論文のタイトルは帰還雷撃の工学モデルです．著者および講演者は南京都大学の電子太郎教授です．

〈補足〉 "I would like to thank Ms. Denki for her nice presentation." といった後，拍手を送ります (applaud by clapping hands)．最後の文は，"The author is Professor Taro Denshi of the University of Southern Kyoto, Japan, and he will also present the paper." ということもできます．

質疑応答の制限時間を超えても質問が続いている場合には，座長は次のようにいって，次の講演に移ります．

> I'm sorry but time is running out. I would like to stay on schedule. So, please continue discussion during the next coffee break. Thank you, Ms. Denki, for your nice presentation. Now, let's move on to the next talk.

> すみませんが，もう時間がございません．スケジュールどおりに進めたいと思いますので，次の休憩時間にディスカッションを続けてください．電気さん，素晴らしいご講演をありがとうございました．それでは，次の講演に移りましょう．

講演時間が超過して質疑応答の時間がない場合には，座長は次のようにいって，次の講演に移ります．

> I'm afraid there is no time for any questions. Let's move on to the next presentation.
>
> すみませんが，質疑応答の時間がございません．次の講演に移りましょう．

次の講演がキャンセルになった場合には，座長は次のようにいいます．

> The next presentation has been cancelled. Let's skip it, and move on to the third paper presentation.
>
> 次の講演はキャンセルになりましたので，3番目の論文の講演に移りましょう．

〈補足〉 事前連絡なしで会場に現れない講演者(no-show)については，対応はむずかしいですが，事前に連絡があった場合には，座長は，たとえば，次のように対応することができます．その論文の題目，著者名および所属を載せたスライドを1枚，その論文のなかで最も興味深いと思われるグラフを載せたスライドを1枚，その論文の結論を載せたスライドを1枚準備しておきます．1枚目の題目などを載せたスライドを見せた後，2枚目のグラフを載せたスライドを見せ，"I think this graph is interesting. The vertical axis represents..., and the horizontal axis represents..."といって，そのグラフがなんであるかを示し，最後に3枚目の結論を載せたスライドを見せます．質疑応答は行わず，"If you have any questions, please contact the first author. His (or her) e-mail address is shown in the paper."といって締めくくります．

2.5.5 セッション締めくくりのことば　*How to Close a Session*

すべての講演と質疑応答が終わったら，座長は次のようにいって，セッションを締めくくります．

> If there are no further questions, I would like to thank Professor Denshi for his excellent presentation. Now, we have completed the presentation of all scheduled papers. At this point, I would like to close this session. Let's thank all the speakers and those who joined the discussions.
>
> ほかにご質問がなければ，素晴らしいご講演に対して電子教授に感謝したいと思います．これで，予定されていたすべての論文発表が終了いたしましたので，本セッションを閉じたいと思います．講演者の皆様とディスカッションに参加してくださった皆様に感謝の意を表しましょう．

〈補足〉「大きな拍手を送りましょう．」は "Let's give a big round of applause." といいます．

セッション参加者に連絡事項がある場合には，座長は次のようにいいます．

> I have two announcements. First is about the starting time for the next session. The next session will start at eleven twenty am. Second is about the banquet. The banquet will start in Hall A on the third floor of this hotel at seven pm this evening.
>
> 皆様に二つ連絡事項がございます．一つ目は次のセッションの開始時間です．次のセッションは午前11：20に開始いたします．二つ目は今晩のバンケットの件です．バンケットは，午後7時から，このホテルの3階のホールAで開かれます．

2.6 発表用ポスターの作成法と発表法
How to Prepare a Poster for a Poster Presentation and How to Talk with the Poster

本節では,発表用ポスターの作成法とそれを用いた発表法について説明します.ポスター発表会場には,通常,A0サイズ(縦1.19メートル,横0.84メートル,面積1平方メートル)またはそれより大きなサイズの板が用意されています.その板に作成したポスターを貼り付け,その傍らに立ち,前に立ち止った方々に研究発表を行います.

2.6.1 ポスターの作成 *How to Prepare a Poster*

ポスターは,数メートル離れたところからでも十分に読める大きさの文字や図を用いて作成しましょう.また,ポスターを見ただけでも研究内容や成果の要点がわかるように工夫して作成しましょう.1枚刷りのポスターの例を図2.13に示します.このポスターは白黒ですが,色使いを工夫して,重要な部分を際立たせましょう.図を横に並べると目をジグザグに動かす必要があるため,図は縦に並べて表示するほうがよいといわれています.

2.6.2 概要の説明 *How to Explain the Outline of a Paper*

ポスターの前で立ち止まり,たとえば,次のように説明を依頼された場合,3〜4分で,研究内容や成果の概要を説明しましょう.

> Hello. I am interested in this research. Could you please explain it for me?
>
> こんにちは.この研究に興味があります.説明していただけませんか.

説明を依頼されなかった場合でも,ポスターの前で立ち止まった方には,次のように声をかけてみて,回答が"Yes"であれば,説明を開始しましょう.

FDTD 2021, Kyoto   Apr. 1, 2021

Influence of Semiconducting Layers on Propagation of Power Line Communication Signals in Power Cables

Hanako Denki and Taro Denshi
University of Southern Kyoto, Kyoto, Japan
hanako.denki@*****.ac.jp

1. Background and Objective

Power line communication (PLC) systems use power lines and cables as data communication media in a frequency range up to **30 MHz**. Power cables have **semiconducting layers**, and are not designed for transmitting PLC signals.

The objective of this study is to reveal the propagation characteristics of a PLC signal along a power cable from the finite-difference time-domain (FDTD) simulation.

2. Model of a Power Cable

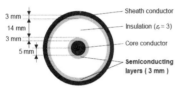

Relative permittivity $\varepsilon_r = 3$
Conductivity $\sigma = 10^{-5}$ to 10^5 S/m

3. Analyzed Results

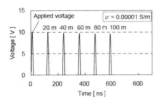

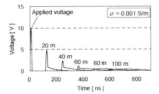

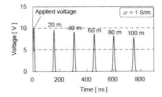

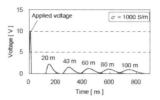

4. Discussion

A. Attenuation and dispersion around $\sigma = 0.001$ S/m

The time constant of the semiconducting layers ($CR = \varepsilon_0 \varepsilon_r / \sigma = 27$ ns) is close to the half cycle (17 ns) of 30-MHz signal. Therefore, the attenuation and dispersion is due to **capacitive charging and discharging of the layers**.

B. Attenuation and dispersion around $\sigma = 1000$ S/m

The penetration depth for $\sigma = 1000$ S/m and $f = 30$ MHz is 2 mm. This is close to the semiconducting-layer thickness (3 mm). Therefore, the attenuation and dispersion is due to **axial current propagation in the semiconducting layers**.

C. Propagation velocity 135 m/μs around $\sigma = 1$ S/m

Since the penetration depth for $\sigma = 1$ S/m and $f = 30$ MHz is 65 mm (>> 3 mm), axial current does not much propagate in the semi-conducting layers. The corresponding inductance is $L' = \mu_0 / 2\pi \cdot \ln(25\,\text{mm}/5\,\text{mm})$.

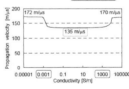

Since the time constant of the semiconducting layers is 0.027 ns, charge moves immediately across the layer. The corresponding capacitance is

$$C' = 2\pi\varepsilon_0\varepsilon_r / \ln\left[(25-3)\,\text{mm}/(5+3)\,\text{mm}\right].$$

From L' and C', the theoretical propagation velocity is $v' = 1/\sqrt{L'C'} = 132$ m/μs. The FDTD-computed value (135 m/μs) agrees well with this theoretical value.

5. Summary

A 30-MHz PLC signal suffers significant attenuation and dispersion when $\sigma = 0.001$ and 1000 S/m. Therefore, it will be difficult to conduct PLC signals in a power system cable with $\sigma = 0.001$ or 1000 S/m, but there are more possibilities if $\sigma \leq 10^{-5}$ S/m or $\sigma = 1$ S/m.

図2.13 ポスターの例

第 2 章　英語論文口頭発表法

> Hello. If you are interested in this research, I would be pleased to explain it for you.
>
> こんにちは．もし，この研究にご興味がございましたら，説明させていただきます．

図 2.13 のポスターを用いた場合の説明を以下に示します．

> My name is Hanako Denki. I am a Ph.D. student of the University of Southern Kyoto. If you have any questions, please stop me at any time.
>
> The objective of this study is to reveal the propagation characteristics of a 30-megahertz power line communication (PLC) signal along a power cable having semiconducting layers using the finite-difference time-domain (FDTD) method.
>
> What you can see here is the cross-section of the model of a power cable that we used in this work. One 3-millimeter-thick semiconductor is installed between the core conductor and the insulating layer, and another 3-millimeter-thick semiconductor is installed between the insulating layer and the sheath conductor. We set its conductivity to a value in a wide range from ten microsiemens per meter up to one hundred thousand siemens per meter, and fixed the relative permittivity to 3.
>
> These four figures show FDTD-computed waveforms of a 30-megahertz signal at different distances from the excitation point along the cable. This figure shows the results computed for a semiconducting-layer conductivity of ten microsiemens per meter. The vertical axis represents the voltage between the core and sheath conductors in volts. The horizontal axis represents time in nanoseconds. The figure shows the results for a semiconducting-layer conductivity of one millisiemen per meter, this shows the results for a conductivity of one siemen per meter, and this shows the results for a conductivity of one thousand siemens per meter.
>
> From these results, you can see the signal magnitude decreases with increasing propagation distance. The attenuation and dispersion are significant when the semiconducting-layer conductivity is about one

millisiemen per meter and about one thousand siemens per meter. Also from these results, you can see the propagation velocity of the signal depends on the semiconducting-layer conductivity. When the semiconducting-layer conductivity is one siemen per meter, the velocity is 135 meters per microsecond. It is lower than those computed for other values of semiconducting-layer conductivity.

Next, I would like to discuss the reason for the high attenuation and dispersion when the semiconducting-layer conductivity is around one millisiemen per meter. They are due to capacitive charging and discharging in a radial direction of the semiconducting layers since the layers have a time constant close to the half cycle of a 30-megahertz signal.

The significant attenuation and dispersion when the semiconducting-layer conductivity is around one thousand siemens per meter are due to axial current propagation in the semiconducting layers. This is because the penetration depth for a medium of conductivity of one thousand siemens per meter and frequency of 30 megahertz is close to the semiconducting-layer thickness.

I would like to skip the discussion on the propagation velocity, and move on to the conclusion. We found a 30-megahertz PLC signal suffers significant attenuation and dispersion when the semiconducting-layer conductivity is around one millisiemen per meter and one thousand siemens per meter. Therefore, it will be difficult to conduct PLC signals in a power system cable with semiconducting layers of conductivity one millisiemen per meter or one thousand siemens per meter, but there are more possibilities if a semiconducting-layer conductivity lower than ten to the minus five siemens per meter or of about one siemen per meter is used.

Thank you very much for your kind attention.

電気花子と申します．南京都大学の博士課程の学生です．もしも質問がございましたら，いつでも止めてください．

本研究の目的は，FDTD 法により，半導電層を有する電力ケーブル上での 30 MHz の信号の伝搬特性を明らかにすることです．

こちらは本研究で使用した電力ケーブルモデルの断面図です．中心導体と絶縁層の間および絶縁層とシース導体の間には厚さ 3 mm の半導電

層が設けられています．半導電層の導電率は 10 μS/m から 100 000 S/m までの広い範囲に設定し，半導電層の比誘電率は 3 一定としました．

これらの四つの図は，電圧印加点から離れたいくつかの点における 30 MHz の信号の FDTD 計算波形を示しています．この図は半導電層の導電率が 10 μS/m の場合の計算結果を示しています．縦軸は，中心導体とシース導体間電圧をボルト単位で示しています．横軸は時間をナノ秒単位で示しています．この図は半導電層の導電率が 1 mS/m の場合の結果を，この図は半導電層の導電率が 1 S/m の場合の結果を，この図は半導電層の導電率が 1 000 S/m の場合の結果を示しています．

これらの結果から，信号の振幅が伝搬に伴って減衰していることがわかると思います．減衰と変わいは，半導電層の導電率が 1 mS/m および 1 000 S/m の場合に顕著となっています．これらの結果から，伝搬速度も半導電層の導電率に依存していることに気づかれたと思います．半導電層の導電率が 1 S/m の場合には，伝搬速度は 135 m/μs となっており，ほかの導電率の場合に比べて伝搬速度が低くなっています．

次に，半導電層の導電率が 1 mS/m 程度の場合に生じる著しい減衰と変わいの原因について検討を行います．それらは半導電層の半径方向における充放電に起因しています．なぜなら半導電層は 30 MHz 信号の半サイクルに近い時定数をもっているからです．

半導電層の導電率が 1 000 S/m の場合に生じる著しい減衰と変わいは，軸方向電流が半導電層を流れることに起因したものであると考えられます．なぜなら，導電率 1 000 S/m，周波数 30 MHz での透過深さが半導電層の厚みに近いからです．

伝搬速度についての検討は省略し，結論に移りたいと思います．周波数 30 MHz の PLC 信号は半導電層の導電率が 1 mS/m および 1 000 S/m 程度のときに，著しい減衰と変わいを生じることがわかりました．したがって，1 mS/m または 1 000 S/m の半導電層を有する電力ケーブルでは，PLC 信号の伝導は困難となりますが，半導電層の導電率が 10^{-5} S/m より低い場合あるいは 1 S/m 程度の場合には，PLC 信号伝導の可能性はより高まるという結論になります．

ご清聴ありがとうございました．

概要説明が終わったら，次のようにいって，コメントをいただきましょう．

> Do you have any questions or comments?
>
> ご質問またはコメントはございませんか．

聴いてくれた方に，例えば，次のようにいって，コメントや意見を促してみることもできます．

> Have you made field measurements of traveling wave response in power cables?
>
> 電力ケーブルで進行波応答の実験を行われたことはございますか．

> If we lower the conductivity that much, do you think it will affect the electrical performance of the cable and joints?
>
> もしも，半導電層の導電率をそこまで下げると，ケーブルやその接続部の電気的特性に影響すると思われますか．

論文のハードコピーあるいは研究内容や成果をより詳しく記述したプリント(handout)を数部用意しておき，興味をもって聴いてくれた方々に名刺代りに配っている発表者もいます．渡す際には，たとえば，次のようにいいましょう．

> I have several hardcopies of this paper（または handouts）here. Please take one.
>
> ここに本論文を印刷したもの(または配布資料)がございます．一部お持ちください．

そして，最後に次のようにお礼をいいましょう．

> Thank you for stopping by.
>
> お立ち寄りくださいまして，ありがとうございました．

2.7 オンライン発表の心得とオンデマンド発表資料の作成
Advice about Online Oral Presentation and Preparation of On-demand Oral Presentation Material

2.7.1 オンライン発表の心得
Advice about Online Oral Presentation

　数年間にわたる新型コロナウイルス感染症の流行により，ウェブ会議ツールが普及しオンライン形式の国際会議も増えてきました．オンライン形式の国際会議には，どこからでも参加できるという大きな利点があります．一方，発表者にとっては，視聴者の表情が見えにくく，反応を見ながら説明速度を変えたり補足説明を行ったりすることが難しいという欠点があります．また，視聴者にとっては，集中力を保って視聴し続けることが難しいという欠点があります．さらに，Wi-Fi などの通信環境が悪い場合，映像や音声が途切れたり，慣れていないウェブ会議ツールの操作に発表者が手間取ると，視聴者の集中力を低下させてしまうという欠点もあります．

　オンライン形式であっても，発表用スライドの構成や説明法等は，2.1 から 2.4 で説明したものと変わりありませんが，各セクションの説明が終わる際に，発表の現在時点と進行状況を確認する目的で，図 2.3 のようなスライドを挿入して示すと，視聴者が集中力を保持して最後まで発表を視聴してくれる可能性が高まると思います．また，Microsoft 社の PowerPoint を使用して発表者側の画面を共有して発表する場合には，アニメーション機能を活用しスライドに動きをもたせ，またスライド画面上の該当部分をマウスポインタで指し示しましょう．さらに，禁止されていない場合にはカメラをオンにして，顔の表情を見せながら発表すると，視聴者に臨場感を与えることができると思います．

　また，発表者にとっては，座長の顔だけでも映してもらえると，視聴者の代表としての反応が見えるので，安心して発表を行うことができます．このように発表者と座長だけでも映せば，発表者と視聴者の間のコミュニケーションの質が高まると思います．

　無線 LAN では近くの電子レンジの使用時に通信速度が遅くなったり途切れたりすることがありますが，有線 LAN を用いれば安定に通信が行えます．

このため，有線LANが利用できる場合には，それをパソコンに接続して使用したほうが良いでしょう．また，充電式のマイクロホンやイヤホンを使用する場合には，自身の発表のタイミングで充電切れが発生しないように準備しておく必要があります．

参加予定の国際会議で使用されるウェブ会議ツールについては，その国際会議の事務局等から事前にお知らせがあり，その使用マニュアルも公開されます．普段使い慣れていないツールであれば，使用マニュアルをよく読み，発表当日も，セッション前の休憩時間等に動作チェックを行っておくのが良いと思います．

2.7.2 オンデマンド発表資料の作成
Preparation of On-demand Oral Presentation Material

オンライン形式の国際会議で，通信の不具合などによりオンライン発表ができない場合に配信する目的で，音声入りの発表動画の提出が求められる場合があります．また，開催時間が深夜になる国々も発生するため，音声入りの発表動画を一定期間ウェブに置き，オンデマンド配信される場合もあります．

PowerPointは，このような音声入り発表用スライド資料の作成にも対応しています．発表用スライドの作成は，対面発表やオンライン発表の場合と変わりませんが，音声を入力(録音)する作業と動画に変換する作業が追加で必要になります．以下では，作成したスライドへの音声入力と動画変換の手順を説明します．

作成したスライドをPowerPointで開き，その編修画面上部にある「スライドショー」メニューを選択します．そうすると，その下に幾つかのメニューが現れます．その中から，「スライドショーの記録」を選択すると，その下にプルダウンメニューが現れます．その中から，「先頭から記録」を選択すると，音声入力の準備が完了です．この時に現れる画面の左上には，「記録」，「停止」，「再生」などのボタンがあり，右下には「カメラ」と「マイク」のオンオフボタンがあります．また，「ポインタオプション」で「レーザポインタ」を選択しておくと，スライド画面上でのマウスポインタの動きも記録されます．最後のスライドまで記録が終了したら，「停止」を押し，ファイルを保存してください．入力した音声やカーソルの動きを変更したい場合には，そのスライドの編修画面に移動し，「スライドショーの記録」のプルダウンメニューの中から「現在のスライドから記録」を選択してください．

次に，PowerPoint編修画面の左上の「ファイル」メニューを選択すると，

左側に幾つかのメニューが現れます．この中から，「エクスポート」を選択し，その次に「ビデオの作成」を選択してください．そして，「フル HD」，「HD」などから希望する解像度を選択し，その下の「記録されたタイミングとナレーションを使用する」が選択されていることを確認し，最後に「ビデオの作成」ボタンを押してください．そうすると，「名前を付けて保存」のポップアップ画面が現れるので，そこに「ファイル名」を入力し，また「ファイルの種類」としては mp4 形式を選択して「保存」を押してください．そうすると，mp4 形式の動画ファイルへの変換が開始します．

第 3 章

電気電子工学分野における諸表現

Various Expressions in Electrical and Electronics Engineering Areas

3.1 電気磁気学における表現
3.2 電気回路理論における表現
3.3 電力・エネルギー分野における表現
3.4 電子デバイスに関する表現
3.5 情報通信に関する表現
3.6 機械学習・人工知能(AI)に関する表現

3.1 電気磁気学における表現
Expressions Used in Electromagnetics

本節では，電気磁気学で用いられる諸表現を示します．

3.1.1 ベクトル解析で用いられる表現
Expressions Used in Vector Analysis

ここでは，スカラ積，ベクトル積，スカラ関数の勾配，ベクトルの流束に関する例文を示します．

> The scalar product between two vectors $A \cdot B$ results in a scalar and is defined as
> $$A \cdot B = AB \cos \theta$$
> where θ is the smallest angle between the two vectors（M. Zahn, Electromagnetic Field Theory: a problem solving approach, p. 11, ©1979 John Wiley & Sons, Inc. Markus Zahn 教授による許諾を得て転載）．
>
> 二つのベクトルのスカラ積 $A \cdot B$ はスカラになり，
> $$A \cdot B = AB \cos \theta$$
> と定義されます．ここで，θ は，これら二つのベクトルの劣角です．

〈補足〉「スカラ積（内積）」を "dot product" または "inner product" と書く場合もあります．

> The vector product between two vectors $A \times B$ results in a vector perpendicular to both vectors in the direction given by the right-hand rule. The magnitude of the vector product is
> $$|A \times B| = AB \sin \theta$$
> where θ is the enclosed angle between A and B（M. Zahn, Electromagnetic Field Theory: a problem solving approach, p. 13, ©1979 John Wiley & Sons, Inc. Markus Zahn 教授による許諾を得て転載）．
>
> 二つのベクトルのベクトル積 $A \times B$ は，右手の法則で与えられる方向

で両ベクトルに垂直なベクトルになります．このベクトル積の大きさは，
$$|\boldsymbol{A}\times\boldsymbol{B}|=AB\sin\theta$$
となります．ここで，θは，ベクトル$\boldsymbol{A}$と$\boldsymbol{B}$の間の内角です．

〈補足〉 "perpendicular to..." は「...に垂直(直角)に」です．なお，「...に平行に」は "parallel to..." と書きます．"vector product"「ベクトル積」を "cross product" または "outer product" と書く場合もあります．

Often we are concerned with the properties of a scalar field $f(x, y, z)$ around a particular point. The chain rule of differentiation gives us the incremental change df in f for a small change in position from (x, y, z) to $(x+\mathrm{d}x, y+\mathrm{d}y, z+\mathrm{d}z)$:

$$\mathrm{d}f = \frac{\partial f}{\partial x}\mathrm{d}x + \frac{\partial f}{\partial y}\mathrm{d}y + \frac{\partial f}{\partial z}\mathrm{d}z \tag{3.1.1.1}$$

If the general differential distance vector $\boldsymbol{dl}$ is defined as
$$\boldsymbol{dl} = \boldsymbol{i}\mathrm{d}x + \boldsymbol{j}\mathrm{d}y + \boldsymbol{k}\mathrm{d}z \tag{3.1.1.2}$$

(3.1.1.1) can be written as the scalar product:

$$\mathrm{d}f = \left(\boldsymbol{i}\frac{\partial f}{\partial x} + \boldsymbol{j}\frac{\partial f}{\partial y} + \boldsymbol{k}\frac{\partial f}{\partial z}\right)\cdot \boldsymbol{dl} = \mathrm{grad}\,f\cdot \boldsymbol{dl} \tag{3.1.1.3}$$

where the spatial derivative terms in parentheses are defined as the gradient of f:

$$\mathrm{grad}\,f = \nabla f = \boldsymbol{i}\frac{\partial f}{\partial x} + \boldsymbol{j}\frac{\partial f}{\partial y} + \boldsymbol{k}\frac{\partial f}{\partial z} \tag{3.1.1.4}$$

The symbol ∇ with the gradient term is introduced as a general vector operator, termed the del operator:

$$\nabla = \boldsymbol{i}\frac{\partial}{\partial x} + \boldsymbol{j}\frac{\partial}{\partial y} + \boldsymbol{k}\frac{\partial}{\partial z} \tag{3.1.1.5}$$

By itself the del operator is meaningless, but when it premultiplies a scalar function, the gradient operator is defined (M. Zahn, Electromagnetic Field Theory: a problem solving approach, p. 16, ©1979 John Wiley & Sons, Inc. Markus Zahn 教授による許諾を得て転載).

私たちは，ある特定の点の周囲におけるスカラ場$f(x, y, z)$の性質に，しばしば関心があります．微分の連鎖法則は，(x, y, z)から$(x+\mathrm{d}x, y+\mathrm{d}y, z+\mathrm{d}z)$への微小な位置変化に対する$f$の漸増変化d$f$を以下のように与えます．

$$df = \frac{\partial f}{\partial x}dx + \frac{\partial f}{\partial y}dy + \frac{\partial f}{\partial z}dz \tag{3.1.1.1}$$

その微分距離ベクトル dl が以下のように定義されると，

$$dl = idx + jdy + kdz \tag{3.1.1.2}$$

(3.1.1.1)式は，以下のようなスカラ積として表されます．

$$df = \left(i\frac{\partial f}{\partial x} + j\frac{\partial f}{\partial y} + k\frac{\partial f}{\partial z}\right) \cdot dl = \operatorname{grad} f \cdot dl \tag{3.1.1.3}$$

ここで，丸括弧内の空間微分項は f の勾配として定義されます．

$$\operatorname{grad} f = \nabla f = i\frac{\partial f}{\partial x} + j\frac{\partial f}{\partial y} + k\frac{\partial f}{\partial z} \tag{3.1.1.4}$$

この勾配項の記号 ∇ は，ベクトル微分演算子と呼ばれる一般的なベクトル演算子の一つとして導入されています．

$$\nabla = i\frac{\partial}{\partial x} + j\frac{\partial}{\partial y} + k\frac{\partial}{\partial z} \tag{3.1.1.5}$$

ベクトル微分演算子は，それ自身では無意味ですが，スカラ関数に左から掛けると，その関数の勾配演算子が定義されます．

〈補足〉 "be concerned with..." は「...に関心がある（関係している）」，"be defined as..." は「...と定義される」，"in parentheses" は「丸括弧内の」，"chain rule of differentiation" は「微分の連鎖法則（律）」，"spatial derivative" は「空間微分」，"gradient of f" は「f の勾配」，"premultiply" は「左から掛ける」，"del operator" は「ベクトル微分演算子 ∇」です．なお，ベクトル演算子記号 ∇ は，"nabla" と読まれることは少なく，ほとんどの場合 "del" と読まれます．また，$\partial f/\partial t$ は「ダイエフ ダイティ」と発音されます．一方，df/dt は「ディエフ ディティ」と発音されます．

We are illustrating with a fluid analogy the flux Φ of a vector A through a closed surface:

$$\Phi = \oint_S A \cdot dS \tag{3.1.1.6}$$

The differential surface element dS is a vector that has magnitude equal to an incremental area on the surface but points in the direction of the outgoing unit normal n to the surface S. Only the component of A perpendicular to the surface contributes to the flux, as the tangential component only results in flow of the vector A along the surface and not through it. A positive contribution to the flux occurs if A has a component in the direction of dS out from the surface. If the normal component of A points into the volume, we have a negative contribution to the flux.

If there is no source for A within the volume V enclosed by the surface S, all the flux entering the volume equals that leaving and the net flux is zero. A source of A within the volume generates more flux leaving than entering so that the flux is positive ($\Phi>0$) while a sink has more flux entering than leaving so that $\Phi<0$ (M. Zahn, Electromagnetic Field Theory: a problem solving approach, p. 22, ©1979 John Wiley & Sons, Inc. Markus Zahn 教授による許諾を得て転載).

ある閉曲面を通り抜けるベクトル A の流束を流体にたとえて説明します．

$$\Phi = \oint_S A \cdot dS \tag{3.1.1.6}$$

微分表面要素 dS は，その表面積に等しい大きさをもち，表面 S に対して外向きの単位法線 n の方向を指すベクトルです．その表面に接する方向の(A の)成分は，表面に沿ったベクトル A の流れとなり表面を通り抜けることはないため，表面に垂直な A の成分のみがその流束に寄与します．A が，その表面から流出する方向の成分をもっている場合，その流束に対して正の寄与が生じます．A の垂直成分が閉曲面で囲まれる空間内を向いている場合には，その流束に対して負の寄与が生じます．

表面 S により囲まれる体積 V 内に A の源がない場合，その体積に流入する全流束は，そこから流出する全流束に等しくなり，正味の流束は零となります．その体積内の A の源が，そこに流入してくるよりも多くの流束を発生すると，流束は正となります($\Phi>0$)．一方，流入してくる流束のほうが多い場合には，$\Phi<0$ となります．

〈補足〉 "unit normal" は「単位法線」，"normal component" は「法線方向成分」，"tangential component" は「接線方向成分」です．なお，「単位法線ベクトル」は "unit normal vector"，「単位接線ベクトル」は "unit tangent vector" と書きます．

3.1.2 電界と磁界に関する表現
Expressions Used for Electric and Magnetic Fields

ここでは，静電界，ストークスの定理，電位差，磁束密度，ベクトルポテンシャル，ヘルムホルツの定理に関する例文を示します．

> If the charge q_1 exists alone, it exerts no force. If we now bring charge q_2 in the vicinity of q_1, then q_1 exerts a force on q_2 that varies in magnitude and direction as it is moved about in space and is thus a way of mapping out the vector force field due to q_1. A charge other than q_2 would feel a different force from q_2 proportional to its own magnitude and sign. It becomes convenient to work with the quantity of force per unit charge that is called the electric field, because this quantity is independent of the particular value of charge used in mapping the force field. Considering q_2 as the test charge, the electric field due to q_1 at the position of q_2 is defined as
>
> $$\boldsymbol{E}_2 = \lim_{q_2 \to 0} \frac{\boldsymbol{F}_2}{q_2} = \frac{q_1}{4\pi\varepsilon_0 r_{12}^2} \frac{\boldsymbol{r}_{12}}{r_{12}} \quad [\text{V/m}] \quad (3.1.2.1)$$
>
> In the definition of (3.1.2.1) the charge q_1 must remain stationary. This requires that the test charge q_2 be negligibly small so that its force on q_1 does not cause q_1 to move (M. Zahn, Electromagnetic Field Theory: a problem solving approach, pp. 56-57, ⓒ1979 John Wiley & Sons, Inc. Markus Zahn 教授による許諾を得て転載).
>
> 電荷 q_1 が単独で存在している場合，その電荷はなんの力も及ぼしません．次に，q_1 の近くに電荷 q_2 をもってくると，q_1 は，q_2 の位置により大きさや方向が変化する力を q_2 に及ぼします．したがって，q_2 は，q_1 によるベクトル力の場をマッピングする一つの手段となります．q_2 以外の電荷が存在すれば，それ自身の大きさ(電荷量)と符号に比例する別の力を q_2 から受けます．電界と呼ばれる単位電荷あたりの力の量を用いれば便利です．なぜなら，電界は力の場をマッピングする際に用いられる特定の電荷量と無関係になるからです．試験電荷として q_2 を考えると，q_2 の位置での q_1 に起因した電界は次のように定義されます．
>
> $$\boldsymbol{E}_2 = \lim_{q_2 \to 0} \frac{\boldsymbol{F}_2}{q_2} = \frac{q_1}{4\pi\varepsilon_0 r_{12}^2} \frac{\boldsymbol{r}_{12}}{r_{12}} \quad [\text{V/m}] \quad (3.1.2.1)$$
>
> 式(3.1.2.1)の定義において，電荷 q_1 は静止していなければなりませ

ん．このためには，試験電荷 q_2 が，q_1 を動かすことのないぐらいに無視できるほど小さくなければなりません．

〈補足〉 "in the vicinity of ..." は「...の近くに」，"be independent of ..." は「...と無関係である」です．なお，「...に依存している」は "be dependent on ..." と書きます．

最後の文の that 節は仮定法現在で，動詞の原形(be)が用いられています．仮定法現在は，"It is necessary" や "require"，"suggest" などに続く必要性や提案などの内容を表す that 節で用いられます．

Using Stokes' theorem, we can convert the line integral of the electric field to a surface integral of the curl of the electric field:

$$\oint_L \boldsymbol{E} \cdot \mathrm{d}\boldsymbol{l} = \int_S (\nabla \times \boldsymbol{E}) \cdot \mathrm{d}\boldsymbol{S} \tag{3.1.2.2}$$

(M. Zahn, Electromagnetic Field Theory: a problem solving approach, pp. 85-86, ©1979 John Wiley & Sons, Inc. Markus Zahn 教授による許諾を得て転載).

ストークスの定理を用いると，電界の線積分を電界の回転の表面積分に変換することができます．

$$\oint_L \boldsymbol{E} \cdot \mathrm{d}\boldsymbol{l} = \int_S (\nabla \times \boldsymbol{E}) \cdot \mathrm{d}\boldsymbol{S} \tag{3.1.2.2}$$

〈補足〉 この文は分詞構文で，"If we use Stokes' theorem, we can convert..." を意味しています．したがって，"using" は分詞ですが，手段や方法を示す前置詞(by や with)のように用いられています．

"line integral" は「線積分」，"surface integral" は「面積分」です．なお，「体積積分」は "volume integral" と書きます．

"Stokes' theorem"「ストークスの定理」のように，人名の語尾にアポストロフィ(エス)をつけ所有格で用いる場合には，固有名詞扱いとなり冠詞をつけません．

The potential difference between the two points at r_a and r_b is the work per unit charge necessary to move from r_a to r_b:

$$V(r_b) - V(r_a) = \frac{W}{q_1} = -\int_{r_a}^{r_b} \boldsymbol{E} \cdot \mathrm{d}\boldsymbol{l} \tag{3.1.2.3}$$

(M. Zahn, Electromagnetic Field Theory: a problem solving approach, p. 86, ©1979 John Wiley & Sons, Inc. Markus Zahn 教授による許諾を得て転載).

r_a と r_b 間の電位差は，単位電荷を r_a から r_b に移動させるのに要する仕事です．

$$V(r_b) - V(r_a) = \frac{W}{q_1} = -\int_{r_a}^{r_b} \boldsymbol{E} \cdot \mathrm{d}\boldsymbol{l} \tag{3.1.2.3}$$

Since the divergence of the magnetic field is zero, we may write the magnetic field as the curl of a vector,

$$\nabla \cdot \boldsymbol{B} = 0 \quad \Rightarrow \quad \boldsymbol{B} = \nabla \times \boldsymbol{A} \tag{3.1.2.4}$$

where A is called the vector potential, as the divergence of the curl of any vector is always zero. Often it is easier to calculate A and then obtain the magnetic field from (3.1.2.4).

From Ampere's law, the vector potential is related to the current density as

$$\nabla \times \boldsymbol{B} = \nabla \times (\nabla \times \boldsymbol{A}) = \nabla(\nabla \cdot \boldsymbol{A}) - \nabla^2 \boldsymbol{A} = \mu_0 \boldsymbol{J} \tag{3.1.2.5}$$

We see that (3.1.2.4) does not uniquely define A, as we can add the gradient of any term to A and not change the value of the magnetic field, since the curl of the gradient of any function is always zero:

$$\boldsymbol{A} \to \boldsymbol{A} + \nabla f \quad \Rightarrow \quad \boldsymbol{B} = \nabla \times (\boldsymbol{A} + \nabla f) = \nabla \times \boldsymbol{A} \tag{3.1.2.6}$$

Helmholtz's theorem states that to uniquely specify a vector, both its curl and divergence must be specified and that far from the sources, the fields must approach zero. To prove this theorem, let's say that we are given the curl and divergence of A and we are to determine what A is. Is there any other vector C, different from A that has the same curl and divergence? We try C of the form

$$\boldsymbol{C} = \boldsymbol{A} + \boldsymbol{a} \tag{3.1.2.7}$$

and we will prove that $\boldsymbol{a}$ is zero.

By definition, the curl of C must equal the curl of A so that the curl of $\boldsymbol{a}$ must be zero:

$$\nabla \times \boldsymbol{C} = \nabla \times (\boldsymbol{A} + \boldsymbol{a}) = \nabla \times \boldsymbol{A} \quad \Rightarrow \quad \nabla \times \boldsymbol{a} = 0 \tag{3.1.2.8}$$

This requires that $\boldsymbol{a}$ can be derived from the gradient of a scalar function f:

$$\nabla \times \boldsymbol{a} = 0 \quad \Rightarrow \quad \boldsymbol{a} = \nabla f \tag{3.1.2.9}$$

Similarly, the divergence condition requires that the divergence of $\boldsymbol{a}$ be zero.

$$\nabla \cdot \boldsymbol{C} = \nabla \cdot (\boldsymbol{A} + \boldsymbol{a}) = \nabla \cdot \boldsymbol{A} \quad \Rightarrow \quad \nabla \cdot \boldsymbol{a} = 0 \tag{3.1.2.10}$$

So that the Laplacian of f must be zero,

$$\nabla \cdot \boldsymbol{a} = \nabla \cdot \nabla f = \nabla^2 f = 0 \tag{3.1.2.11}$$

In the preceding chapter, we obtained a similar equation and solution for the electric potential that goes to zero far from the charge distribution:

$$\nabla^2 V = -\frac{\rho}{\varepsilon} \quad \Rightarrow \quad V = \int_V \frac{\rho \, \mathrm{d}V}{4\pi\varepsilon r} \tag{3.1.2.12}$$

If we equate f to V, then ρ must be zero giving us that the scalar function f is also zero. That is, the solution of Laplace's equation of (3.1.2.11) for zero sources everywhere is zero, even though Laplace's equation in a region does have nonzero solutions if there are sources in other regions of space. With f zero, from (3.1.2.9) we have that the vector $\boldsymbol{a}$ is also zero and then $\boldsymbol{C} = \boldsymbol{A}$, thereby proving Helmholtz's theorem (M. Zahn, Electromagnetic Field Theory: a problem solving approach, pp. 336-338, ©1979 John Wiley & Sons, Inc. Markus Zahn 教授による許諾を得て転載).

　磁界の発散は零であり，いかなるベクトルであっても，その回転の発散はいつも零となるため，あるベクトルの回転として磁界を表現できます．

$$\nabla \cdot \boldsymbol{B} = 0 \quad \Rightarrow \quad \boldsymbol{B} = \nabla \times \boldsymbol{A} \tag{3.1.2.4}$$

ここで，$\boldsymbol{A}$ はベクトルポテンシャルと呼ばれています．多くの場合，$\boldsymbol{A}$ を計算し，それから式(3.1.2.4)により磁界を求めることのほうが簡単です．

　アンペアの法則より，ベクトルポテンシャルは次のように電流密度と関係しています．

$$\nabla \times \boldsymbol{B} = \nabla \times (\nabla \times \boldsymbol{A}) = \nabla(\nabla \cdot \boldsymbol{A}) - \nabla^2 \boldsymbol{A} = \mu_0 \boldsymbol{J} \tag{3.1.2.5}$$

　いかなる関数であっても，その勾配の回転はいつも零となるため，いかなる項の勾配を $\boldsymbol{A}$ に加えても磁界の値は変化しません．このため，式(3.1.2.4)は $\boldsymbol{A}$ を一意的には定義していないことがわかります．

$$\boldsymbol{A} \rightarrow \boldsymbol{A} + \nabla f \quad \Rightarrow \quad \boldsymbol{B} = \nabla \times (\boldsymbol{A} + \nabla f) = \nabla \times \boldsymbol{A} \tag{3.1.2.6}$$

ヘルムホルツの定理は，あるベクトルを一意的に定めるためには，そ

のベクトルの回転と発散の両方が定められる必要があり，源から遠く離れたところでは場が零に近づく必要があることを述べています．この定理を証明するため，A の回転と発散が与えられているとし，A を決めてみましょう．A 以外で，A と同じ回転と発散をもつベクトル C はあるでしょうか？　次の形の C を試し，

$$C = A + a \tag{3.1.2.7}$$

a が零になることを証明します．

　定義により，C の回転は A の回転に等しくなければならないため，a の回転は零にならなければなりません．

$$\nabla \times C = \nabla \times (A + a) = \nabla \times A \quad \Rightarrow \quad \nabla \times a = 0 \tag{3.1.2.8}$$

このためには，a があるスカラ関数 f の勾配の形となる必要があります．

$$\nabla \times a = 0 \quad \Rightarrow \quad a = \nabla f \tag{3.1.2.9}$$

同様に，発散の条件より，a の発散は零である必要があります．

$$\nabla \cdot C = \nabla \cdot (A + a) = \nabla \cdot A \quad \Rightarrow \quad \nabla \cdot a = 0 \tag{3.1.2.10}$$

したがって，f のラプラシアンは零である必要があります．

$$\nabla \cdot a = \nabla \cdot \nabla f = \nabla^2 f = 0 \tag{3.1.2.11}$$

前章において，電荷が分布しているところから離れたところでは零に近づく電位についての同じような方程式と解を得ました．

$$\nabla^2 V = -\frac{\rho}{\varepsilon} \quad \Rightarrow \quad V = \int_v \frac{\rho \, dV}{4\pi \varepsilon r} \tag{3.1.2.12}$$

f が V に等しいと仮定すると，ρ が零でなければならず，スカラ関数 f もまた零となります．つまり，空間内のほかの場所に源があり，ある部分におけるラプラス方程式が零ではない解をもつとしても，あらゆる場所で源が零である場合の式(3.1.2.11)のラプラス方程式の解は零となるということです．f が零であれば，式(3.1.2.9)より，ベクトル a もまた零となり，$C = A$ となります．これにより，ヘルムホルツの定理が証明されます．

〈補足〉 "the divergence of a vector" は「ベクトルの発散」，"the curl of a vector" は「ベクトルの回転」，"the gradient of a scalar function" は「スカラ関数の勾配」，"vector potential" は「ベクトルポテンシャル」，"current density" は「電流密度」，"Ampere's law" は「アンペアの法則」，"Helmholtz's theorem" は「ヘルムホルツの定理」，"Laplacian" は「ラプラシアン」，"Laplace's equation" は「ラプラス方程式」，"be related to..." は「...と関連(関係)している」，"by definition" は「定義により(よれば)」です．

　式(3.1.2.9)の下の文の that 節は仮定法現在であるため，動詞の原形(be)

3.1.3 電磁波に関する表現
Expressions Used for Electromagnetic Waves

ここでは，ファラデーの法則，アンペアの法則，ガウスの法則，マクスウェル方程式，変位電流，電荷保存則，ポインティングベクトルに関する例文を示します．

In the historical development of electromagnetic field theory through the nineteenth century, charge and its electric field were studied separately from currents and their magnetic fields. Until Faraday showed that a time varying magnetic field generates an electric field, it was thought that the electric and magnetic fields were distinct and uncoupled. Faraday believed in the duality that a time varying electric field should also generate a magnetic field, but he was not able to prove this supposition.

It remained for James Clerk Maxwell to show that Faraday's hypothesis was correct and that without this correction Ampere's law and conservation of charge were inconsistent:

$$\nabla \times \boldsymbol{H} = \boldsymbol{J} \quad \Rightarrow \quad \nabla \cdot \boldsymbol{J} = 0 \tag{3.1.3.1}$$

$$\nabla \cdot \boldsymbol{J} + \frac{\partial \rho}{\partial t} = 0 \tag{3.1.3.2}$$

for if we take the divergence of Ampere's law in (3.1.3.1), the current density must have zero divergence because the divergence of the curl of a vector is always zero. This result contradicts (3.1.3.2) if a time varying charge is present. Maxwell realized that adding the displacement current on the right-hand side of Ampere's law would satisfy charge conservation, because of Gauss's law relating $\boldsymbol{D}$ to ρ ($\nabla \cdot \boldsymbol{D} = \rho$).

This simple correction has far-reaching consequences, because we will be able to show the existence of electromagnetic waves that travel at the speed of light c, thus proving that light is an electromagnetic wave. Because of the significance of Maxwell's correction, the complete set of coupled electromagnetic field laws is called Maxwell's equations:
Faraday's Law

$$\nabla \times \boldsymbol{E} = -\frac{\partial \boldsymbol{B}}{\partial t} \quad \Rightarrow \quad \oint_L \boldsymbol{E} \cdot \mathrm{d}\boldsymbol{l} = -\frac{\mathrm{d}}{\mathrm{d}t}\int_S \boldsymbol{B} \cdot \mathrm{d}\boldsymbol{S} \qquad (3.1.3.3)$$

Ampere's law with Maxwell's displacement current correction

$$\nabla \times \boldsymbol{H} = \boldsymbol{J} + \frac{\partial \boldsymbol{D}}{\partial t} \quad \Rightarrow \quad \oint_L \boldsymbol{H} \cdot \mathrm{d}\boldsymbol{l} = \int_S \boldsymbol{J} \cdot \mathrm{d}\boldsymbol{S} + \frac{\mathrm{d}}{\mathrm{d}t}\int_S \boldsymbol{D} \cdot \mathrm{d}\boldsymbol{S}$$
$$(3.1.3.4)$$

Gauss's laws

$$\nabla \cdot \boldsymbol{D} = \rho \quad \Rightarrow \quad \oint_S \boldsymbol{D} \cdot \mathrm{d}\boldsymbol{S} = \int_V \rho \, \mathrm{d}V \qquad (3.1.3.5)$$

$$\nabla \cdot \boldsymbol{B} = 0 \quad \Rightarrow \quad \oint_S \boldsymbol{B} \cdot \mathrm{d}\boldsymbol{S} = 0 \qquad (3.1.3.6)$$

Conservation of charge

$$\nabla \cdot \boldsymbol{J} + \frac{\partial \rho}{\partial t} = 0 \quad \Rightarrow \quad \oint_S \boldsymbol{J} \cdot \mathrm{d}\boldsymbol{S} + \frac{\mathrm{d}}{\mathrm{d}t}\int_V \rho \, \mathrm{d}V = 0 \qquad (3.1.3.7)$$

As we have justified, (3.1.3.7) is derived from the divergence of (3.1.3.4) using (3.1.3.5).

Note that (3.1.3.6) is not independent of (3.1.3.3) for if we take the divergence of Faraday's law, $\nabla \cdot \boldsymbol{B}$ could at most be a time independent function. Since we assume that at some point in time $\boldsymbol{B}=0$, this function must be zero.

The symmetry in Maxwell's equations would be complete if a magnetic charge density appeared on the right-hand side of Gauss's law in (3.1.3.6) with an associated magnetic current due to the flow of magnetic charge appearing on the right-hand side of (3.1.3.3). Thus far, no one has found a magnetic charge or current, although many people are actively looking. Throughout this text we accept (3.1.3.3)–(3.1.3.7) keeping in mind that if magnetic charge is discovered, we must modify (3.1.3.3) and (3.1.3.6) and add an equation like (3.1.3.7) for conservation of magnetic charge (M. Zahn, Electromagnetic Field Theory: a problem solving approach, pp. 488–489, ©1979 John Wiley & Sons, Inc. Markus Zahn 教授による許諾を得て転載).

19世紀の電磁界理論の歴史的発展のなかで，電流と磁界と切り離して電荷と電界は研究されました．ファラデーが時間変化する磁界が電界を発生することを示すまでは，電界と磁界は別々の関連のないものと考えられていました．ファラデーは，時間変化する電界もまた磁界を発生

するはずであるという双対性を信じていましたが，この仮定を証明することはできませんでした．

　ファラデーの仮説が正しく，この修正がなければアンペアの法則と電荷保存則が矛盾することを示したのは，ジェイムズ・クラーク・マクスウェルでした．

$$\nabla \times \boldsymbol{H} = \boldsymbol{J} \quad \Rightarrow \quad \nabla \cdot \boldsymbol{J} = 0 \tag{3.1.3.1}$$

$$\nabla \cdot \boldsymbol{J} + \frac{\partial \rho}{\partial t} = 0 \tag{3.1.3.2}$$

というのは，式(3.1.3.1)のアンペアの法則の発散をとると，ベクトルの回転の発散はいつも零となるため，電流密度の発散が零とならなければならないからです．この結果は，時間変化する電荷が存在する場合，式(3.1.3.2)に矛盾します．マクスウェルは，$\boldsymbol{D}$ と ρ を関連づけるガウスの法則($\nabla \cdot \boldsymbol{D} = \rho$)により，アンペアの法則の右辺に変位電流を加えることで電荷保存則を満たせることに気づいたのです．

　この単純な修正は広範囲にわたる影響を及ぼします．なぜなら，光速 c で伝搬する電磁波の存在を示すことができ，それにより光が電磁波であることを証明することができるからです．マクスウェルの修正の重要性のため，以下に示す密接な関係にある電磁界の法則一式はマクスウェル方程式と呼ばれています．

ファラデーの法則

$$\nabla \times \boldsymbol{E} = -\frac{\partial \boldsymbol{B}}{\partial t} \quad \Rightarrow \quad \oint_L \boldsymbol{E} \cdot \mathrm{d}\boldsymbol{l} = -\frac{\mathrm{d}}{\mathrm{d}t} \int_S \boldsymbol{B} \cdot \mathrm{d}\boldsymbol{S} \tag{3.1.3.3}$$

マクスウェルによる変位電流の修正を含むアンペアの法則

$$\nabla \times \boldsymbol{H} = \boldsymbol{J} + \frac{\partial \boldsymbol{D}}{\partial t} \quad \Rightarrow \quad \oint_L \boldsymbol{H} \cdot \mathrm{d}\boldsymbol{l} = \int_S \boldsymbol{J} \cdot \mathrm{d}\boldsymbol{S} + \frac{\mathrm{d}}{\mathrm{d}t} \int_S \boldsymbol{D} \cdot \mathrm{d}\boldsymbol{S} \tag{3.1.3.4}$$

ガウスの法則

$$\nabla \cdot \boldsymbol{D} = \rho \quad \Rightarrow \quad \oint_S \boldsymbol{D} \cdot \mathrm{d}\boldsymbol{S} = \int_V \rho \, \mathrm{d}V \tag{3.1.3.5}$$

$$\nabla \cdot \boldsymbol{B} = 0 \quad \Rightarrow \quad \oint_S \boldsymbol{B} \cdot \mathrm{d}\boldsymbol{S} = 0 \tag{3.1.3.6}$$

電荷保存則

$$\nabla \cdot \boldsymbol{J} + \frac{\partial \rho}{\partial t} = 0 \quad \Rightarrow \quad \oint_S \boldsymbol{J} \cdot \mathrm{d}\boldsymbol{S} + \frac{\mathrm{d}}{\mathrm{d}t} \int_V \rho \, \mathrm{d}V = 0 \tag{3.1.3.7}$$

すでに証明したように，式(3.1.3.7)は，式(3.1.3.4)の発散と式(3.1.3.5)から導かれます．

> なお，式(3.1.3.6)は式(3.1.3.3)と独立ではありません．というのは，ファラデーの法則の発散をとると，$\nabla \cdot B$ はよくても時間に依存しない関数にしかなりえないためです．私たちは，ある時点で $B=0$ を仮定しますので，この関数は零にならなければなりません．
>
> もしも磁荷密度が式(3.1.3.6)のガウスの法則の右辺にあり，式(3.1.3.3)の右辺に磁荷の流れに起因した磁流があれば，マクスウェル方程式の対称性は完全になるでしょう．多くの人が今も活発に探していますが，これまでのところ，誰も磁荷あるいは磁流を発見するにはいたっていません．この教科書では，もしも磁荷が発見されたら，式(3.1.3.3)と式(3.1.3.6)を修正し，磁荷保存則のための式(3.1.3.7)のような式を追加しなければならないことを心に留めて，式(3.1.3.3)から式(3.1.3.7)を受け入れています．

〈補足〉 "time varying magnetic field" は「時間変化する磁界」，"time varying electric field" は「時間変化する電界」，"hypothesis" は「仮説」，"conservation of charge" または "charge conservation" は「電荷保存」，"displacement current" は「変位電流」，"right-hand side" は「右側，(数式の)右辺」，"speed of light" は「光速」です．

第1段落の第2文 "Until Faraday showed that a time varying magnetic field generates..." において，主節の動詞が過去形(showed)であるのに従属節の動詞が現在形(generates)になっているのは，従属節が現在においても正しい「不変の真理」を示しているためです(時制の一致の例外)．

最後の段落の第1文 "The symmetry in Maxwell's equations would be complete if a magnetic charge density appeared on..." は仮定法過去の文で，現在の事実に反する仮定を述べています．

> We expand the vector quantity
>
> $$\nabla \cdot (E \times H) = H \cdot (\nabla \times E) - E \cdot (\nabla \times H) = -H \cdot \frac{\partial B}{\partial t} - E \cdot \frac{\partial D}{\partial t} - E \cdot J$$
>
> (3.1.3.8)
>
> where we change the curl terms using Faraday's and Ampere's laws.
>
> For linear homogeneous media, including free space, the constitutive laws are
>
> $$D = \varepsilon E, \quad B = \mu H \qquad (3.1.3.9)$$
>
> so that (3.1.3.8) can be rewritten as

$$\nabla \cdot (\boldsymbol{E} \times \boldsymbol{H}) + \frac{\partial}{\partial t}\left(\frac{1}{2}\varepsilon E^2 + \frac{1}{2}\mu H^2\right) = -\boldsymbol{E} \cdot \boldsymbol{J} \tag{3.1.3.10}$$

which is known as Poynting's theorem. We integrate (3.1.3.10) over a closed volume, using the divergence theorem to convert the first term to a surface integral:

$$\oint_S (\boldsymbol{E} \times \boldsymbol{H}) \cdot \mathrm{d}\boldsymbol{S} + \frac{\mathrm{d}}{\mathrm{d}t}\int_V \left(\frac{1}{2}\varepsilon E^2 + \frac{1}{2}\mu H^2\right)\mathrm{d}V = -\int_V \boldsymbol{E} \cdot \boldsymbol{J}\,\mathrm{d}V \tag{3.1.3.11}$$

We recognize the time derivative in (3.1.3.11) as operating on the electric and magnetic energy densities, which suggests the interpretation of (3.1.3.11) as

$$P_{\mathrm{out}} + \frac{\mathrm{d}W}{\mathrm{d}t} = -P_d \tag{3.1.3.12}$$

where P_{out} is the total electromagnetic power flowing out of the volume with density

$$\boldsymbol{S} = \boldsymbol{E} \times \boldsymbol{H} \quad [\mathrm{W/m^2}] \tag{3.1.3.13}$$

where S is called the Poynting vector, W is the electromagnetic stored energy, and P_d is the power dissipated or generated:

$$\left.\begin{aligned} P_{\mathrm{out}} &= \oint_S (\boldsymbol{E} \times \boldsymbol{H}) \cdot \mathrm{d}\boldsymbol{S} = \oint_S \boldsymbol{S} \cdot \mathrm{d}\boldsymbol{S} \\ W &= \int_V \left(\frac{1}{2}\varepsilon E^2 + \frac{1}{2}\mu H^2\right)\mathrm{d}V \\ P_d &= \oint_V \boldsymbol{E} \cdot \boldsymbol{J}\,\mathrm{d}V \end{aligned}\right\} \tag{3.1.3.14}$$

If $\boldsymbol{E}$ and $\boldsymbol{J}$ are in the same direction as in an Ohmic conductor ($\boldsymbol{E} \cdot \boldsymbol{J} = \sigma E^2$), then P_d is positive, representing power dissipation since the right-hand side of (3.1.3.12) is negative. A source that supplies power to the volume has $\boldsymbol{E}$ and $\boldsymbol{J}$ in opposite directions so that P_d is negative (M. Zahn, Electromagnetic Field Theory: a problem solving approach, pp. 490-491, ©1979 John Wiley & Sons, Inc. Markus Zahn 教授による許諾を得て転載).

ベクトル量を次のように展開します.

$$\begin{aligned}\nabla \cdot (\boldsymbol{E} \times \boldsymbol{H}) &= \boldsymbol{H} \cdot (\nabla \times \boldsymbol{E}) - \boldsymbol{E} \cdot (\nabla \times \boldsymbol{H}) \\ &= -\boldsymbol{H} \cdot \frac{\partial \boldsymbol{B}}{\partial t} - \boldsymbol{E} \cdot \frac{\partial \boldsymbol{D}}{\partial t} - \boldsymbol{E} \cdot \boldsymbol{J}\end{aligned} \tag{3.1.3.8}$$

ここで，ファラデーの法則とアンペアの法則を用いて回転の項を変えています．

自由空間を含む線形均質な媒質では，構成則は次のようになります．
$$D=\varepsilon E, \quad B=\mu H \tag{3.1.3.9}$$
その結果，式(3.1.3.8)は次のように書き換えられます．
$$\nabla \cdot (E \times H) + \frac{\partial}{\partial t}\left(\frac{1}{2}\varepsilon E^2 + \frac{1}{2}\mu H^2\right) = -E \cdot J \tag{3.1.3.10}$$
これはポインティングの定理として知られています．閉空間で式(3.1.3.10)を(体積)積分し，発散の定理を用いて第1項を表面積分に変換すると，次式が得られます．
$$\oint_S (E \times H) \cdot dS + \frac{d}{dt}\int_V \left(\frac{1}{2}\varepsilon E^2 + \frac{1}{2}\mu H^2\right)dV = -\int_V E \cdot J\, dV \tag{3.1.3.11}$$

式(3.1.3.11)の時間微分は，電界と磁界のエネルギー密度に作用していることがわかります．このことは，式(3.1.3.11)を次のように解釈できることを示唆しています．
$$P_{\text{out}} + \frac{dW}{dt} = -P_d \tag{3.1.3.12}$$
ここで，P_{out} は閉空間から流出する電磁界の仕事率で，次のような密度で表されます．
$$S = E \times H \quad [\text{W/m}^2] \tag{3.1.3.13}$$
ここで，S はポインティングベクトル，W は電磁蓄積エネルギー，P_d は消費あるいは発生電力です．
$$\left. \begin{aligned} P_{\text{out}} &= \oint_S (E \times H) \cdot dS = \oint_S S \cdot dS \\ W &= \int_V \left(\frac{1}{2}\varepsilon E^2 + \frac{1}{2}\mu H^2\right)dV \\ P_d &= \oint_V E \cdot J\, dV \end{aligned} \right\} \tag{3.1.3.14}$$
オームの法則に従う導体中のように E と J が同じ方向であれば($E \cdot J = \sigma E^2$)，P_d は正となり，式(3.1.3.12)の右辺が負となるため，電力は消費されます．その閉空間に電力を供給する電源は，反対方向の E と J をもつため，P_d は負となります．

〈補足〉 "expand" は「(式を)展開する」，"homogeneous media (medium の複数形)" は「均質な媒質」，"free space" は「自由空間」，"electric and magnetic energy densities" は「電界と磁界のエネルギー密度」, "Poynting vector" は「ポインティングベクトル」です.

3.2 電気回路理論における表現
Expressions Used in Electrical Circuit Theory

本節では，電気回路理論で用いられる諸表現を示します．

3.2.1 直流回路と交流回路に関する表現
Expressions Used for Direct-current and Alternating-current Circuits

ここでは，オームの法則，抵抗，キャパシタンス，インダクタンス，インピーダンス，力率，共振，電力，星形結線，渦電流，キルヒホッフの法則，テブナンの定理に関する例文を示します．

Ohm's law states that the current I flowing in a circuit is directly proportional to the applied voltage V and inversely proportional to the resistance R, provided the temperature remains constant. Thus,

$$I = \frac{V}{R} \quad \text{or} \quad V = IR \quad \text{or} \quad R = \frac{V}{I}$$

(J. Bird, Electrical Circuit Theory and Technology, p. 14, ©2003 Newnes. Taylor & Francis Books UK による許諾を得て転載).

オームの法則は，温度が一定であれば，ある回路に流入する電流 I は印加電圧 V に比例し，その回路の抵抗 R に反比例することを述べています．したがって，次のように表されます．

$$I = \frac{V}{R} \quad \text{or} \quad V = IR \quad \text{or} \quad R = \frac{V}{I}$$

〈補足〉 "current flow in ..." は「...に流入する電流」, "applied voltage" は「印加電圧」, "be directly proportional to ..." は「...に比例する」, "be inversely proportional to ..." は「...に反比例する」, "provided ..." は「もしも ... であれば」です．

In general, as the temperature of a material increases, most conductors increase in resistance, insulators decrease in resistance, while the resistances of some special alloys remain almost constant.

The temperature coefficient of resistance of a material is the increase in the resistance of a 1 Ω resistor of that material when it is subjected to a rise of temperature of 1 K. The symbol used for the temperature coefficient of resistance is α. Thus, if a copper wire of resistance 1 Ω is heated through 1 K and its resistance is then measured as 1.0043 Ω then $\alpha=0.0043$ Ω/ΩK for copper. The unit is usually expressed only as "per K", i.e., $\alpha=0.0043$/K for copper. If the 1 Ω resistor of copper at a reference temprature of 20 ℃ is heated to 100 ℃, the resistance at the temperature would be 1+80 K×0.0043/K=1.344 Ω (J. Bird, Electrical Circuit Theory and Technology, p. 26, ©2003 Newnes. Taylor & Francis Books UK による許諾を得て転載).

一般に，物質の温度が上昇するにつれて，ほとんどの導体の抵抗は上昇し，絶縁体の抵抗は低下します．一方，特殊な合金のなかには，ほぼ一定の抵抗を示すものもあります．

ある物質の抵抗の温度係数とは，その物質でできた 1 Ω の抵抗器が 1 K の温度上昇を受けたときの抵抗上昇のことです．抵抗の温度係数に用いられる記号は α です．したがって，1 Ω の銅線が 1 K 熱せられ，抵抗が 1.0043 Ω になったとすると，銅では $\alpha=0.0043$ Ω/ΩK となります．単位は通常は「1/K」のみ，すなわち，銅では $\alpha=0.0043$/K で表されます．基準温度 20 ℃ での 1 Ω の銅の抵抗器が 100 ℃ まで熱せられたとすると，その温度でのその抵抗は，1+80 K×0.0043/K=1.344 Ω となります．

〈補足〉 "i.e." はラテン語 "id est" の略語で，論文中でよく使う略語です（1.1.13 の表 1.2 参照）．これは英語の "that is" に対応し，「すなわち」，「いい換えれば」を意味します．"be subjected to ..." は「... を受ける（さらされる）」または「... による」です．

Every system of electrical conductors possesses capacitance. For example, there is capacitance between the conductors of overhead transmission lines and also between the wires of a telephone cable. In these examples, the capacitance is undesirable but has to be accepted, minimized or compensated for. There are other situations where capacitance is a desirable property.

Devices specially constructed to possess capacitance are called capacitors (or condensers, as they used to be called). In its simplest form a capacitor consists of two plates which are separated by an insulating material known as a dielectric. A capacitor has the ability to store a quantity of static electricity (J. Bird, Electrical Circuit Theory and Technology, p. 57, ©2003 Newnes. Taylor & Francis Books UK による許諾を得て転載).

すべての導体系はキャパシタンスをもっています．たとえば，架空送電線の導体間や電話ケーブルの導線間にキャパシタンスがあります．これらの例では，キャパシタンスは望ましくはありませんが，受け入れられるか，最小化されるか，あるいは補償されなければなりません．一方，キャパシタンスが望ましい性質となるような状況もあります．

キャパシタンスをもつように特別に作られた素子はキャパシタと呼ばれています（かつてはコンデンサと呼ばれていました）．最も単純な形態では，キャパシタは，誘電体として知られている絶縁物質により隔てられた二つの板から構成されます．キャパシタは静電気量を蓄える能力があります．

〈補足〉 "overhead (power) transmission line" は「架空送電線」，"dielectric" は「誘電体」です．なお，「架空配電線」は "overhead (power) distribution line" と書きます．

Apparent power, S, is an important quantity since alternating-current apparatus, such as generators, transformers and cables, is usually rated in voltamperes rather than in watts. The allowable output of such apparatus is usually limited not by mechanical stress but by temperature rise, and hence by the losses in the device. The losses are determined by the voltage and current and are almost independent of the power factor. Thus the amount of electrical equipment installed to supply a certain load is essentially determined by the voltamperes of the load rather than by the power alone. The rating of a machine is defined as the maximum apparent power that it is designed to carry continuously without overheating.

The reactive power, Q, contributes nothing to the net energy transfer and yet it causes just as much loading of the equipment as real power of the same magnitude. Reactive power is a term used widely in power generation, distribution and utilization of electrical energy (J. Bird, Electrical Circuit Theory and Technology, p. 464, ©2003 Newnes. Taylor & Francis Books UK による許諾を得て転載).

発電機，変圧器，ケーブルのような交流機器は通常ワットではなくボルトアンペアで定格が決められるため，皮相電力 S は重要な量です．このような機器の許容出力は，通常機械的応力だけではなく温度上昇，したがって，その機器の損失によっても制限されます．損失は電圧と電流で決まり，力率にはほぼ無関係です．上述したように，ある負荷に電力を供給するために設置される電気機器の容量は，基本的には，電力のみよりは負荷のボルトアンペアにより決められます．機器の定格は，過熱することなく連続的に送ることができるように設計された最大皮相電力として定義されます．

無効電力 Q は，正味のエネルギー伝達には全く寄与しませんが，同じ大きさの有効電力のように，その機器の負荷となります．無効電力は，発電，配電および電気エネルギー利用分野において広く用いられる語です．

〈補足〉 "apparent power" は「皮相電力」，"alternating-current (a.c.)" は「交流(の)」，"power factor" は「力率」，"apparatus" は「機器，装置」，"equipment" は「機器，装置」，"rating" は「定格」，"reactive power" は「無効電力」です．

A 30 μF capacitor is connected in parallel with an 80 Ω resistor across a 240 V, 50 Hz supply. Calculate (a) the current in each branch, (b) the supply current, (c) the circuit phase angle, (d) the circuit impedance, (e) the power dissipated, and (f) the apparent power (J. Bird, Electrical Circuit Theory and Technology, p. 240, ©2003 Newnes. Taylor & Francis Books UK による許諾を得て転載).

240 V, 50 Hz の電源に接続された 80 Ω の抵抗器に，30 μF のキャパシタが並列に接続されています．(a) 各枝の電流，(b) 電源電流，(c)

回路の位相角，(d) 回路のインピーダンス，(e) 消費電力，および (f) 皮相電力を計算しなさい．

〈補足〉 "in parallel with..." は「...に並列に」，"phase angle" は「位相角」です．なお，「消費電力」は "dissipated power" または "consumed power"，「有効電力」は "active power" と書きます．有効電力の単位は "watts (W)"，無効電力の単位は "voltamperes reactive (VAR)"，皮相電力の単位は "voltamperes (VA)" です．

A coil of negligible resistance and inductance 100 mH is connected in series with a capacitance of 2 μF and a resistance of 10 Ω across a 50 V, variable frequency supply. Determine (a) the resonant frequency, (b) the current at resonance, (c) the voltages across the coil and the capacitor at resonance, and (d) the Q-factor of the circuit (J. Bird, Electrical Circuit Theory and Technology, p. 228, ©2003 Newnes. Taylor & Francis Books UK による許諾を得て転載).

50 V で可変周波数の電源に接続された 2 μF のキャパシタンスと 10 Ω の抵抗に，抵抗が無視でき，インダクタンスが 100 mH のコイルが直列に接続されています．(a) 共振周波数，(b) 共振時の電流，(c) 共振時のコイル，キャパシタ両端の電圧，および (d) 回路の Q 値を求めなさい．

〈補足〉 "in series with..." は「...に直列に」，"variable frequency" は「可変周波数」，"resonant frequency" は「共振周波数」，"Q-factor" は「Q 値」です．

A series circuit possesses resistance R and capacitance C. The circuit dissipates a power of 1.732 kW and has a power factor of 0.866 leading. If the applied voltage is given by $v = 141.4 \sin(10^4 t + \pi/9)$ volts, determine (a) the current flowing and its phase, (b) the value of resistance R, and (c) the value of capacitance C (J. Bird, Electrical Circuit Theory and Technology, p. 468, ©2003 Newnes. Taylor & Francis Books UK による許諾を得て転載).

ある直列回路が抵抗 R とキャパシタンス C をもっています．この回路は 1.732 kW の電力を消費し，進み力率 0.866 をもっています．印加

第3章 電気電子工学分野における諸表現

電圧が $v=141.4\sin(10^4 t+\pi/9)$ V である場合，(a) 回路に流れる電流とその位相，(b) 抵抗 R の値，および (c) キャパシタンス C の値を求めなさい．

〈補足〉 "leading" は「進みの」を意味します．なお，「遅れの」は "lagging" と書きます．

A star-connected load consists of three identical coils each of resistance 30 Ω and inductance 127.3 mH. If the line current is 5.08 A, calculate the line voltage if the supply frequency is 50 Hz (J. Bird, Electrical Circuit Theory and Technology, p. 299, ©2003 Newnes. Taylor & Francis Books UK による許諾を得て転載)．

ある星形接続された負荷は，同一のコイル三つで構成されています．各コイルは抵抗 30 Ω，インダクタンス 127.3 mH をもっています．線電流が 5.08 A であり，電源周波数が 50 Hz である場合の線間電圧を計算しなさい．

〈補足〉 "star-connected" は「星形接続された」，"consist of..." は「...で構成される」，"line current" は「線電流」，"line voltage" は「線間電圧」です．なお，「デルタ接続された」は "delta-connected" と書きます．

If a coil is wound on a ferromagnetic core (such as in a transformer) and alternating current is passed through the coil, an alternating flux is set up in the core. The alternating flux induces an electromotive force e in the coil given by $e=N(\mathrm{d}\phi/\mathrm{d}t)$. However, in addition to the desirable effect of inducing an electromotive force in the coil, the alternating flux induces undesirable voltages in the iron core. These induced electromotive forces set up circulating currents in the core, known as eddy currents. Since the core possesses resistance, the eddy currents heat the core, and this represents wasted energy (J. Bird, Electrical Circuit Theory and Technology, p. 696, ©2003 Newnes. Taylor & Francis Books UK による許諾を得て転載)．

ある強磁性体コア(変圧器内にあるような)にコイルが巻かれており，そのコイルに交流電流が流れると，そのコア内には交流磁束が作られま

> す．この交流磁束は，コイルに $e=N(\mathrm{d}\phi/\mathrm{d}t)$ で与えられる起電力を誘起します．しかし，コイルに起電力を誘起するという望ましい効果に加えて，鉄芯に望ましくない電圧を誘起します．これらの誘導起電力は，渦電流として知られる循環電流をコアに流します．コアも抵抗をもっているため，渦電流はコアを熱します．このことは，エネルギーの消耗を意味します．

〈補足〉 "wound" は「巻かれた」("wind"「巻く」の過去，過去分詞形)，"ferromagnetic core" は「強磁性体コア」，"transformer" は「変圧器」，"alternating flux" は「交流(または交番)磁界」，"electromotive force" は「起電力」，"in addition to…" は「…に加えて」，"iron core" は「鉄芯」，"circulating current" は「循環電流」，"eddy current" は「渦電流」です．なお，「反磁性の」は "diamagnetic" と書きます．

> Kirchhoff's laws may be stated as:
> (a) At any point in an electrical circuit the phasor sum of the currents flowing towards that junction is equal to the phasor sum of the currents flowing away from the junction.
> (b) In any closed loop in a network, the phasor sum of the voltage drops (i.e., the products of current and impedance) taken around the loop is equal to the phasor sum of the electromotive forces acting in that loop (J. Bird, Electrical Circuit Theory and Technology, p. 531, ©2003 Newnes. Taylor & Francis Books UK による許諾を得て転載).
>
> キルヒホッフの法則は次のように述べられます．
> (a) 電気回路のどの点においても，その接続点に流れ込む電流のベクトル和は，その接続点から流れ出る電流のベクトル和に等しい．
> (b) 回路網のどの閉路においても，その閉路を一周してとった電圧降下(すなわち，電流とインピーダンスの積)のベクトル和はその閉路の起電力のベクトル和に等しい．

〈補足〉 "phasor sum" は「ベクトル和」です．"vector sum" といい換えることもできます．

> Thévenin's theorem is stated as follows:

第3章　電気電子工学分野における諸表現

> The current that flows in any branch of a network is the same as that which would flow in the branch if it were connected across a source of electrical energy, the electromotive force of which is equal to the potential difference which would appear across the branch if it were open-circuited, and the internal impedance of which is equal to the impedance which appears across the open-circuited branch terminals when all sources are replaced by their internal impedances (J. Bird, Electrical Circuit Theory and Technology, p. 574, ©2003 Newnes. Taylor & Francis Books UK による許諾を得て転載).
>
> テブナンの定理は次のように述べられます．
> 　回路網のある枝に流れる電流は，その枝が開放されたときに，その枝両端に現れる電位差に等しい起電力の電気エネルギー源と，回路網内のすべての電源がそれらの内部インピーダンスに置き換えられたときの開放された枝の端子に現れるインピーダンスに等しい内部インピーダンスに，その枝が接続された場合に，そこに流れる電流に等しい．

〈補足〉　"branch" は「枝」，"internal impedance" は「内部インピーダンス」，"open-circuited" は「開放された」です．なお，「短絡された」は "short-circuited" と書きます．

3.2.2　過渡現象に関する表現
Expressions Used for Electrical Transient Phenomena

ここでは，集中定数回路の過渡現象，時定数，伝送線路，フィルタに関する例文を示します．

> When a direct-current (d.c.) voltage is applied to a capacitor C, and a resistor R connected in series, there is a short period of time immediately after the voltage is connected, during which the current flowing in the circuit and voltages across C and R are changing.
> 　Similarly, when a d.c. voltage is connected to a circuit having an inductor L connected in series with a resistor R, there is a short period of time immediately after the voltage is connected, during which the current flowing the circuit and the voltages across L and R are changing (J. Bird, Electrical Circuit Theory and Technology, p. 259, ©2003

抵抗 R に直列に接続されたキャパシタ C に直流電圧が印加されると，その直後に，その回路に流入する電流および C および R に加わる電圧が変化する短い期間がある．

同様に，抵抗 R に直列に接続されたインダクタ L を有する回路に直流電圧が印加されると，その直後に，その回路に流入する電流および L および R に加わる電圧が変化する短い期間がある．

〈補足〉 "direct-current (d.c.)" は「直流（の）」，"a voltage is applied to..." は「ある電圧が...に印加される」，"voltage across..." は「...（の両端）に加わる電圧」，"... is connected to 〜" は「...が〜に接続される」です．

A 20 µF capacitor is connected in series with a 50 kΩ resistor, and the circuit is connected to a 20 V, d.c. supply. Determine (a) the initial value of the current flowing, (b) the time constant of the circuit, (c) the value of the current one second after connection, (d) the value of the capacitor voltage two seconds after connection, and (e) the time after connection when the resistor voltage is 15 V (J. Bird, Electrical Circuit Theory and Technology, p. 265, ©2003 Newnes. Taylor & Francis Books UK による許諾を得て転載).

20 µF のキャパシタが 50 kΩ の抵抗器に直列に接続されており，この回路は 20 V の直流電源に接続されています．(a) 流入電流の初期値，(b) 回路の時定数，(c) 接続後 1 秒経過した後の電流値，(d) 接続後 2 秒経過した後のキャパシタ電圧，および (e) 接続後に抵抗器両端の電圧が 15 V になる時間を求めなさい．

〈補足〉 "time constant" は「時定数」です．

An inductor has a negligible resistance and an inductance of 200 mH, and is connected in series with a 1 kΩ resistor to a 24 V, d.c. supply. Determine the time constant of the circuit and the steady-state of the current flowing in the circuit. Find (a) the current flowing in the circuit at a time equal to one time constant, (b) the voltage drop across the inductor at a time equal to two time constants, and (c) the voltage drop across the resistor after a time equal to three time constants (J. Bird, Electrical Circuit Theory and Technology, p. 274, ©2003 Newnes. Taylor & Francis Books UK による許諾を得て転載).

あるインダクタは無視できる抵抗と 200 mH のインダクタンスをもっており，1 kΩ の抵抗器と 24 V の直流電源に直列に接続されています．この回路の時定数とこの回路に流れる定常電流を求めなさい．また，(a) 時定数に等しい時間における回路電流，(b) 時定数の 2 倍に等しい時間におけるインダクタでの電圧降下，および (c) 時定数の 3 倍に等しい時間が経過した後の抵抗器での電圧降下を求めなさい．

〈補足〉 "steady-state" は「定常状態の」です．「過渡状態の」は "transient" と書きます．なお，"resistance"「抵抗」，"inductance"「インダクタンス」，"capacitance"「キャパシタンス」，"impedance"「インピーダンス」は可算名詞でもありますので，これらは状況に応じて単数形あるいは複数形として使用されます．

A transmission line is a system of conductors connecting one point to another and along which electromagnetic energy can be sent. Thus telephone lines and power distribution lines are typical examples of transmission lines; in electronics, however, the term usually implies a line used for the transmission of radio-frequency energy such as that from a radio transmitter to the antenna.

An important feature of a transmission line is that it should guide energy from a source at the sending end to a load at the receiving end without loss by radiation. One form of construction often used consists of two similar conductors mounted close together at a constant separation. The two conductors form the two sides of a balanced circuit and any radiation from one of them is neutralized by that from the other. Such twin-wire lines are used for carrying high radio-frequency power. The coaxial form of construction is commonly employed for low power use, one conductor being in the form of a cylinder which surrounds the other at its center, and thus acts as a screen. Such cables are often used to couple frequency-modulation and television receivers to their antennas.

At frequencies greater than 1 000 MHz, transmission lines are usually in the form of a waveguide which may be regarded as coaxial lines without the center conductor, the energy being launched into the guide or abstracted from it by probes or loops projecting into the guide (J. Bird, Electrical Circuit Theory and Technology, p. 869, ⓒ2003 Newnes. Taylor & Francis Books UK による許諾を得て転載).

　伝送線路とはある点と別の点を接続する導体系のことで，それに沿って電磁エネルギーが送られます．したがって，電話線や配電線は伝送線路の典型的な例です．しかし，電子工学分野においては，この用語は，無線送信機からアンテナへの伝送のような無線周波数のエネルギー伝送に使われる線路を意味します．

　伝送線路の重要な特長は，放射による損失なしで，送電端の源から受電端の負荷にエネルギーを導くことです．よく用いられる構造の一つは，一定の間隔で近接して取り付けられた2本の同じような導体で構成されます．これらの2導体は平衡回路の2辺を形成し，一方からの放射は他方からの放射により打ち消されます．このような2線導体は，高い無線周波数の電力を送るのに用いられています．同軸構造は，通常，小電力用に用いられています．そして，一方の導体は中央の他導体を囲む円筒の形で，遮へいの機能を果たします．このようなケーブルは，FMやテレビの受信機とそれらのアンテナをつなぐのによく用いられています．

　1 000 MHz を超える周波数においては，伝送線路は，通常，中心導体のない同軸線路とみなされる導波路の形態となります．そして，導波路に突き出たプローブあるいはループにより，エネルギーが導波路に送り出されるか，あるいは取り込まれます．

〈補足〉 "transmission line"は「伝送線路」，"electromagnetic energy"は「電磁エネルギー」，"telephone line"は「電話線」，"power distribution line"は「配電線」，"radio frequency"は「無線周波数」，"radio transmitter"は「無線送信機」，"coaxial"は「同軸の」，"frequency modulation (f.m.)"は「周波数変調」，"waveguide"は「導波路」，"abstract"は「取り込む」です．

　第2パラグラフの第5文 "The coaxial form of construction is ..., one conductor being in the form of ～" および第3パラグラフの文 "transmission lines are usually in the form of ..., the energy being launched ～" は独立分詞構文で，それぞれ，"The coaxial form of construction is ..., and one conductor is in the form of ～", "transmission lines are usually in the form of ..., and the energy is launched ～" を意味しています．

153

第3章　電気電子工学分野における諸表現

A transmission line can be considered to consist of a network of a very large number of cascaded T-sections each a very short length of transmission line. It is an approximation of the uniformly distributed line; the larger the number of lumped parameter sections, the nearer it approaches the true distributed nature of the line. When the generator V_s is connected, a current I_s flows which divides between that flowing through the leakage conductance G, which is lost, and that which progressively charges each capacitor C and which sets up the voltage traveling wave moving along the transmission line. The loss or attenuation in the line is caused by both the conductance G and the series resistance R (J. Bird, Electrical Circuit Theory and Technology, p. 871, ©2003 Newnes. Taylor & Francis Books UK による許諾を得て転載).

　伝送線路は，それぞれが非常に短い伝送線路であるT形区間の非常に多数の縦続接続から構成されると考えることができます．それは，均一な分布線路の近似で，集中パラメータ区分の数が多いほど，その線路の真の分布特性により近づきます．発電機 V_s が接続されると，電流 I_s は，漏れコンダクタンス G に流れ込み失われるものと，各キャパシタ C を漸次充電し，伝送線路に沿って移動する電圧進行波を作り出すものに分かれます．その線路の損失あるいは減衰はコンダクタンス G と直列抵抗 R の両方に起因します．

〈補足〉　"cascaded"は「直列（縦続）の」，"traveling wave"は「進行波」です．

A lossless transmission line has a characteristic impedance of $600\angle 0°$ Ω and is connected to an antenna of impedance $(250+j200)$ Ω. Determine (a) the magnitude of the ratio of the reflected to the incident voltage wave, and (b) the incident voltage if the reflected voltage is $10\angle 60°$ Ω (J. Bird, Electrical Circuit Theory and Technology, p. 959, ©2003 Newnes. Taylor & Francis Books UK による許諾を得て転載).

　ある無損失伝送線路は特性インピーダンス $600\angle 0°$ Ω をもち，それがインピーダンス $(250+j200)$ Ω のアンテナに接続されています．(a) 反射電圧波の入射電圧波に対する比の大きさ，および(b) 反射電圧が10

∠60°Ωの場合の入射電圧を求めなさい．

〈補足〉 "characteristic impedance"は「特性インピーダンス」，"reflected voltage wave"は「反射電圧波」，"incident voltage wave"は「入射電圧波」，"ratio of ... to 〜"「...の〜に対する比」です．

A filter is required to pass all frequencies above 50 kHz and to have a nominal impedance of 620 Ω. Design (a) a high-pass T section filter, and (a) a high-pass π section filter to meet these requirements (J. Bird, Electrical Circuit Theory and Technology, p. 958, ©2003 Newnes. Taylor & Francis Books UK による許諾を得て転載)．

あるフィルタは 50 kHz 以上のすべての周波数を通し，公称インピーダンス 620 Ω をもつことが要求されています．これらの条件を満たす (a) 高域通過 T 形フィルタおよび (b) 高域通過 π 形フィルタを設計しなさい．

〈補足〉 "nominal impedance"は「公称インピーダンス」，"high-pass filter"は「高域通過フィルタ」です．

3.2.3　測定器に関する表現
Expressions for Electrical Measuring Instruments

ここでは，可動コイル形計器，電流計，電圧計，マルチメータ，陰極線オシロスコープに関する例文を示します．

A moving-coil instrument, which measures only direct current, may be used in conjunction with a bridge rectifier circuit to provide an indication of alternating currents and voltages. The average value of the full wave rectified current is $0.637I_m$, where I_m is the maximum (crest) value of the sine-wave current. However, a meter being used to measure alternating current is usually calibrated in root-mean-square (RMS) values. For sinusoidal quantities the indication is $0.707I_m/0.637I_m$ i.e. 1.11 times the mean value. Recifier instruments have scales calibrated in root-mean-square quantities and it is assumed by the manufacturer that the a.c. is sinusoidal (J. Bird, Electrical Circuit Theory and

Technology, p. 114, ©2003 Newnes. Taylor & Francis Books UK による許諾を得て転載).

　直流のみを計測する可動コイル形計器は，ブリッジ整流回路とつなぐことによって，交流電流や交流電圧の指示に用いられます．全波整流電流の平均値は，正弦波最大値がI_mの場合，$0.637I_m$となります．しかし，交流を計測するのに用いられる計器は実効値(RMS)で校正されます．正弦波に対しては，表示は$0.707I_m/0.637I_m$，すなわち平均値の1.11倍となります．整流器を含む計器は実効値で校正された目盛りをもち，製造会社により，交流は正弦波であることが仮定されています．

〈補足〉 "moving-coil instrument"は「可動コイル形計器」，"in conjunction with…"は「…とつないで(とともに)」，"rectifier"は「整流器」，"full wave rectified current"は「全波整流電流」，"sinusoidal"は「正弦関数の」，"calibrate"は「校正(較正)する」，"root-mean-square (RMS)"は「実効値(二乗平均平方根)」です．

　An ammeter, which measures current, has a low resistance (ideally zero) and must be connected in series with the circuit.
　A voltmeter, which measures potential difference (p.d.), has a high resistance (ideally infinite) and must be connected in parallel with the part of the circuit whose p.d. is required.
　There is no difference between the basic moving-coil instrument used to measure current and voltage since both use a milliammeter as their basic part. This is a sensitive instrument which gives full-scale deflection for currents of only a few milliamperes. When an ammeter is required to measure currents of larger magnitude, a proportion of the current is diverted through a low-value resistance connected in parallel with the meter. Such a diverting resistor is called a shunt (J. Bird, Electrical Circuit Theory and Technology, p. 115, ©2003 Newnes. Taylor & Francis Books UKによる許諾を得て転載).

　電流計は，電流を計測するもので，低い抵抗(理想的には零)をもち，回路に直列に接続されなければなりません．
　電圧計は，電位差を計測するもので，高い抵抗(理想的には無限大)をもち，対象とする回路の電位差を計測したい部分に並列に接続されなけ

3.2 電気回路理論における表現

　　ればなりません．
　　　電流計も電圧計も，それらの基本部分としてミリアンペア計を用いていますので，電流および電圧を計測するのに用いられている可動コイル形基本計器に相違はありません．これは，数ミリアンペアの電流で最大の振れとなる感度の高い計器です．より大きな電流を計測する必要があるときには，電流計に並列に接続された低抵抗に電流の一部が流されます．このような分流抵抗器は分流器と呼ばれています．

〈補足〉 "ammeter"は「電流計」，"voltmeter"は「電圧計」，"instrument"は「計器(機器)」，"milliammeter"は「ミリアンペア計」，"deflection"は「(計器の)振れ」，"shunt"は「分流器」です．なお，「検流計」は"galvanometer"と書きます．

　　Instruments are manufactured that combine a moving-coil meter with a number of shunts and series multipliers, to provide a range of readings on a single scale graduated to read current and voltage. If a battery is incorporated then resistance can also be measured. Such instruments are called multimeters. The AVOmeter™ is a typical example. A particular range may be selected either by the use of separate terminals or by a selector switch. Only one measurement can be performed at a time. Often such instruments can be used in alternating-current as well as direct-current circuits when a rectifier is incorporated in the instrument (J. Bird, Electrical Circuit Theory and Technology, p. 118, © 2003 Newnes. Taylor & Francis Books UK による許諾を得て転載).

　　単一目盛りで広範囲の電流および電圧を読むことができるようにするため，可動コイル形計器と多数の分流器と直列倍率器を組み合せた計器が製作されています．電池が組み込まれていれば，抵抗も測定できます．このような計器はマルチメータと呼ばれています．アボメータがその一例です．分離端子を利用するか，あるいは切換えスイッチにより特定の範囲が選択できます．一度に一測定しかできません．その計器に整流器が組み込まれている場合には，直流だけではなく交流の測定にも用いることができます．

〈補足〉 "graduate"は「目盛りをつける」，"battery"は「電池」，"AVOmeter™"は「アボメータ」，"selector switch"は「切換えスイッチ」です．

157

An important feature of the AVOmeter™ and similar universal multimeters is a fast-acting "cutout" that interrupts the circuit when the meter current is too high. When the meter has a high voltage scale, this often uses a separately labeled high voltage terminal. Most high-quality moving-coil meters also feature a mirror behind the needle to eliminate parallax error. The eye is in a correct position for accurate reading when the needle and its mirror image converge.

アボメータや類似する汎用マルチメータの重要な特徴の一つは，測定電流が大きすぎる場合に回路を遮断する高速作動「カットアウト」機能です．マルチメータに高電圧スケールがある場合，通常は，別々にラベル付けされた高電圧端子が備わっています．高品質の可動コイルメータのほとんどには，視差誤差を排除するために針の後ろに鏡が付けられています．針とその鏡像が一致するとき，目は正確な読み取りができる正しい位置にあります．

〈**補足**〉 "parallax error"は「視差誤差」です．

The cathode ray oscilloscope may be used in the observation of waveforms and for the measurement of voltage, current, frequency, phase and periodic time. For examining periodic waveforms the electron beam is deflected horizontally (i.e. in the x direction) by a sawtooth generator acting as a timebase. The signal to be examined is applied to the vertical deflection system (y direction) usually after amplification.

Oscilloscopes normally have a transparent grid of 10 mm by 10 mm squares in front of the screen. Among the timebase controls is a "variable" switch which gives the sweep speed as time per centimeter. This may be in s/cm, ms/cm or μs/cm. Also on the front panel of a cathode ray oscilloscope is a y amplifier switch marked in volts per centimeter (J. Bird, Electrical Circuit Theory and Technology, p. 121, © 2003 Newnes. Taylor & Francis Books UKによる許諾を得て転載).

陰極線オシロスコープは波形の観測や電圧，電流，周波数，位相および周期の測定に用いられます．周期波形を観測するために，時間軸の機能を果たすのこぎり波発生器により電子ビームは水平方向(すなわちx

方向に）に偏向させられます．観測対象の信号は，通常，増幅後，垂直偏向システム（y 方向）に加えられます．

オシロスコープは，そのスクリーンの前に，電子ビームを通過させられる 10 mm×10 mm の正方格子をもっています．時間軸調整器の中に，センチメートルごとの時間として掃引速度を与える「可変」スイッチがあります．これは，s/cm，ms/cm あるいは μs/cm 単位です．また，陰極線オシロスコープの前面操作盤には，センチメートルごとの電圧を単位として記された y 方向増幅スイッチがあります．

〈補足〉 "cathode ray oscilloscope" は「陰極線オシロスコープ」，"electron beam" は「電子ビーム」，"sawtooth generator" は「のこぎり波発生器」，"amplification" は「増幅」，"transparent" は「透明な（透過的な）」です．

Digital instrumentation has replaced moving coil meters and cathode ray oscilloscopes in most applications. A digital multimeter has an internal semiconductor bandgap voltage reference and a series of comparators that form an analog to digital converter system. Its operation is similar to a moving coil AVOmeter™ in that a single measurement, voltage, current or resistance, can be performed based on range switches and terminal connections. A digital oscilloscope has internal semiconductor memory that stores a time sequence of values from the analog to digital converter. The sampling rate and display are controlled by a timebase switch. Digital oscilloscopes can measure and compare two, four or more signals at the same time. Digital display offers cursors which can be adjusted to measure difference in time $\varDelta t$, difference in voltage $\varDelta V$ or voltage at a specific time.

ディジタル計測器は，ほとんどの応用において，可動コイル形計器や陰極線オシロスコープに取って代っています．ディジタルマルチメータは，内部に半導体バンドギャップ電圧基準とアナログ-ディジタル変換システムを形成する一連の比較器をもっています．ディジタル計測器の操作は，電圧，電流あるいは抵抗単独の測定はレンジスイッチと端子接続に基づいて行われるという点で，可動コイル形アボメータに似ています．ディジタルオシロスコープは，内部にアナログ-ディジタル変換器からの時系列値を記憶しておく半導体メモリをもっています．そのサンプリングレートと表示は時間基準スイッチにより制御されます．ディジ

タルオシロスコープは，2，4 あるいはそれ以上の信号を同時に測定および比較することができます．ディジタルディスプレイには，時間の差 Δt，電圧の差 ΔV または特定の時点の電圧を測定するために調整できるカーソルがあります．

〈補足〉 "digital instrumentation" は「ディジタル計測器」，"digital multimeter" は「ディジタルマルチメータ」，"semiconductor bandgap voltage reference" は「半導体バンドギャップ電圧基準」，"analog to digital converter" は「アナログ-ディジタル変換器」，"digital oscilloscope" は「ディジタルオシロスコープ」，"sampling rate" は「サンプリングレート」です．

3.3 電力・エネルギー分野における表現
Expressions Used in Electric Power and Energy Area

本節では，電力・エネルギー分野で用いられる諸表現を示します．

3.3.1 電力システムに関する表現
Expressions for Electric Power Systems

ここでは，スマートグリッド，風力発電および瞬時電圧低下に関する例文を示します．

In countries with mature economies, power distribution networks play a critical role in supporting the industrialized society. These power grids were designed decades ago, with the main aim of delivering electricity from large power stations to households and businesses. The last few years, however, have witnessed the introduction of novel technologies and concepts that promise to change the way we produce, manage, and consume electricity. Technologies such as distributed generation and plug-in hybrid electric vehicles (PHEVs) will help to reduce CO_2 emissions and offer more sustainable options to consumers of energy, while applications such as the advanced metering infrastructure (AMI) and home energy management system (HEMS) will enable consumers to manage their energy usage more efficiently. The smart grid initiative aims at modernizing the current electricity grid by introducing a new set of technologies and services that will make the electricity networks more reliable, efficient, secure, and environmentally friendly.

The smart grid will be characterized by a two-way flow of electricity and information, creating an automated, widely distributed energy delivery network. It will incorporate into the power grid the benefits of distributed computing and communications to deliver real-time information that will help balance power supply and demand. To be successful, the smart grid initiative will require collaboration, integration, and interoperability among an array of technologies and disciplines. Various forms of information and communication technology (ICT) will therefore play a major role in facilitating the realization of the modern power grid and its services.

There is an increasing consensus that the communication infrastructure that supports the operation of the power grid today needs fundamental changes. This communication infrastructure was designed to meet the needs of a regulated power industry that dates from several decades ago. The communication networks in these power grids were designed to support control operations and interactions between control centers and individual substations. These legacy control systems, often referred to as supervisory control and data acquisition (SCADA) systems, were typically built using a star topology in which data are exchanged between the control center and substations; the main aims were to detect faults and manage generation and demand in the power grid. The messages exchanged through the communication networks consist of either information about the health of the power grid (voltages, the temperatures of the cables, the status of circuit breakers, and so on) or commands that enable changes in the configuration of the network, such as those used to open and close circuit breakers. Such communication infrastructures therefore have a limited ability to cope with the new requirements of the smart grid in terms of penetration, scalability, and performance (F. Bouhafs, M. Mackay, and M. Merabti, p. 25, ©2001 IEEE. IEEEによる許諾を得て転載).

経済が成熟した国々では，工業化社会を支えるうえで，電力流通網が

重要な役割を果たしています．これらの電力網は，電力を大規模発電所から家庭や企業に輸送することを主目的とし，数十年前に設計されました．しかし，ここ数年は，電力の生産，管理および消費する方法を変える見込みがある新しい技術や概念が導入されるのを目にしてきました．分散発電やプラグインハイブリッド電気自動車(PHEV)のような技術は，二酸化炭素の排出を削減するのに役立ち，より持続可能な選択肢をエネルギー消費者に提供します．そして，高度計量インフラストラクチャ(AMI)やホームエネルギー管理システム(HEMS)のようなアプリケーションは，消費者による効率的なエネルギー利用の管理を可能にします．スマートグリッド構想は，電力網を，より信頼性高く，より効率的に，より安全に，より環境に優しくする一連の新しい技術とサービスを導入することにより，今の電力網を近代化することを目指しています．

　スマートグリッドは電力と情報の双方向の流れを特徴とし，自動化された，広く分布したエネルギー輸送網を作り出します．それは，電力の供給と需要の均衡を保つのに役立つ実時間情報を送るために，分散コンピューティングと通信の恩恵を電力網に取り入れるでしょう．スマートグリッド構想が成功するためには，多くの技術と分野の間での連携，統合および相互運用を必要とするでしょう．したがって，さまざまな形態の情報通信技術(ICT)は，近代の電力網とそのサービスの実現を促進するうえで重要な役割を果たすでしょう．

　電力網の運用を今日支えている通信基盤は根本的な変化を要するというコンセンサスが高まっています．この通信基盤は，数十年前に始まる規制された電力産業の要求を満たすために設計されました．これらの電力網のなかの通信網は，制御センタと個々の変電所の間の制御操作や交信を支えるために設計されました．しばしば監視制御・データ収集(SCADA)システムと称される，これらの旧来の制御システムは，制御センタと変電所間でデータがやり取りされるスター形接続を用いて通常は作られました．その主目的は電力網内の故障を検出することと発電と需要を管理することでした．それらの通信網を通してやり取りされるメッセージは電力網の健全状態(電圧，ケーブル温度，遮断器の状態など)に関する情報か，あるいは遮断器の開閉に使われるような電力網の構成を変更させる命令からなります．したがって，このような通信基盤には，普及性，拡張性および性能において，スマートグリッドの新しい要求に対応するには，かぎられた能力しかありません．

〈補足〉 "power distribution network" は「電力流通網」，"industrialized

society"は「工業化社会」，"power station"は「発電所」，"distributed generation"は「分散発電」，"plug-in hybrid electric vehicle (PHEV)"は「プラグインハイブリッド電気自動車」，"CO_2 emission"は「二酸化炭素の排出」，"sustainable"は「持続可能な」，"advanced metering infrastructure (AMI)"は「高度計量インフラストラクチャ (AMI)」，"home energy management system (HEMS)"は「ホームエネルギー管理システム (HEMS)」，"smart grid"は「スマートグリッド」，"real-time information"は「実時間情報」，"power supply and demand"は「電力の供給と需要」，"information and communication technology (ICT)"は「情報通信技術 (ICT)」，"substation"は「変電所」，"legacy system"は「旧来のシステム」，"referred to as..."は「...と称される」，"supervisory control and data acquisition (SCADA)"は「監視制御・データ収集 (SCADA)」，"circuit breaker"は「遮断器」，"scalability"は「拡張性」です．

たとえば上記の第3文 "The last few years, however, have witnessed..."「しかし，ここ数年は...を目にしてきました．」は時を主語とする無生物主語構文です．このような無生物主語構文は英文ではよく用いられます．

As mentioned previously, modern wind plants can mimic conventional facilities in many respects, but the supply of fuel is variable and cannot be predicted with perfect accuracy over even the next hour. Situations may arise, however, in a power system where it would be beneficial from a reliability perspective (e.g., to prevent overloading a line under an emergency condition after a parallel line is lost) to reduce wind plant output. Curtailments under periods of transmission congestion or extremely low system loads are other examples.

System frequency is one of the primary measures of the "health" of a large, interconnected electric power system. Frequency gives an indication of the balance between supply and demand: declining frequency indicates more demand than supply, while rising frequency results from more supply than demand. Further, frequency under conditions of balance must be maintained within a tight window, usually within tens of mHz of the target 50 to 60 Hz.

While maintaining the interconnection frequency at the target during "normal" conditions is a feat in and of itself, since demand continuously changes over multiple time scales as the result of millions of individual and automated decisions by end users and end-use equipment, it is the sudden disruption of the tenuous supply-demand balance that brings the potential for the greatest consequences. The sudden loss of one or more generating units due to mechanical failure or the loss of significant transmission system elements (that are importing power into an area) may put system frequency into a temporary "free fall." What happens in the few seconds that follow this development makes the difference between a reliable system and widespread blackouts.

In response to the falling frequency, conventional generating units will give up a portion of their stored kinetic energy (in the rotational energy of the turbine generator shaft) as increased power output, which helps retard the frequency decline. Within a few seconds, governor controls on individual generator units will autonomously increase power input from prime movers, further increasing electrical output. The combined response of the units must be sufficient to first arrest the frequency decline and then act to stabilize frequency and move it back towards the desired value (Zavadil et al., pp. 92-93, ©2011 IEEE. IEEEによる許諾を得て転載).

　前述したように，現在の風力発電所は多くの点で従来形の設備と似ていますが，その燃料供給(風量)は変化しやすく，1時間先でさえ完全な精度で予測することはできません．しかし，信頼性の観点から(たとえば，並列送電線が失われた後の緊急時における別の送電線の過負荷を防ぐため)，そのようにすることが有益な場合には，ある電力系統において，風力発電所の出力を減らす状況が生じるかもしれません．送電混雑時あるいは極低負荷時の切離しは別の例です．
　系統周波数は，大規模連系系統の「健全性」の主要な評価基準の一つです．周波数は需給バランスの指標となります．周波数の低下は需要が供給を上回っていることを示し，周波数の上昇は需要を上回る供給により生じます．さらに，均衡条件下の周波数は，狭い範囲内に，通常は50から60Hzの目標周波数に対し数十ミリHz以内に維持されなければなりません．
　末端使用者や装置による何百万もの個別・自動決定の結果として，複数の時間スケールで需要は常に変化するため，通常の運転状態において連系周波数を目標値に維持することは，それ自体むずかしい技術です．一方，最大の(悪い)結果をもたらす可能性があるのは，不安定な需給バランスの突然の崩壊です．機械的故障による1機あるいはそれ以上の発

電装置の突然の解列，あるいは重要な送電系統要素(電力を地域に送り出している)の損失は，系統周波数を一時的に「急降下」させる可能性があります．この後に続く数秒間に起こることが，信頼性の高い系統と広範囲に及ぶ停電との差となります．

降下する周波数に応答して，従来の発電装置は，それらに蓄積されていた運動エネルギー(タービン発電機軸の回転エネルギーとして)の一部を電力出力の増加として差し出します．それは周波数低下を遅らせるのに役立ちます．数秒以内に，各発電装置の調速機制御装置が自律的に原動機からの入力を増やし，さらに電気的出力を増やします．各発電装置の複合的な応答は，まずは周波数の低下を止め，そして周波数を安定化させ，望ましい値に戻すのに十分である必要があります．

〈補足〉 "wind plant"は「風力発電所」，"power system"は「電力系統」，"reliability perspective"は「信頼性の観点」，"curtailment"は「切離し」，"transmission congestion"は「送電混雑」，"system frequency"は「系統周波数」，"blackout"は「停電」，"in response to..."は「...に応答して」，"kinetic energy"は「運動エネルギー」，"rotational energy"は「回転エネルギー」，"turbine generator shaft"は「タービン発電機軸」，"governor control"は「調速機制御」，"autonomously"は「自律的に」，"prime mover"は「原動機」です．

The Kriegers Flak offshore project in the Baltic Sea is looked upon as the first technical prototype of an offshore grid. Transmission system operators (TSO) in the three neighboring countries of Sweden, Germany and Denmark have carried out feasibility studies to interconnecting an offshore wind power plant from each country by establishing the first offshore node of an offshore grid. Parts of this project are currently under construction in the Baltic Sea.

Many answers to practical questions concerning the market-driven operation of a meshed, multiterminal voltage source converter (VSC) high voltage direct current (HVdc) grid are still missing. Since late 2008, the Danish TSO has been working on an internal project aimed at closing this gap. A simulation benchmark test system has been developed, facilitating detailed investigations on operational questions.

Results have already been published showing that extensive coordination of control systems is essential, both with respect to avoiding unwanted dc loop flows and with respect to finding the location of the optimal slack node in the system (Holttinen et al., pp. 56-57, ©2011 IEEE. IEEE による許諾を得て転載).

　バルチック海におけるクリーガース・フラック洋上事業は，洋上電力系統の最初の技術的プロトタイプとみなされます．スウェーデン，ドイツおよびデンマークの３隣国の送電系統運用者(TSO)は，洋上電力系統の最初のノードを作ることにより，各国の洋上風力発電所を連系できるか否かについて検討してきました．この事業の何割かは，バルチック海において現在建設中です．

　多端子電圧形変換(VSC)直流高電圧(HVdc)の網目状の系統を，市場主導で運用することに関する実際問題に対する多くの答えは今もありません．2008年の後半から，デンマークのTSOは，この隔たりを埋めることを目的とした内部プロジェクトに取り組んできました．シミュレーションベンチマークテストシステムが開発され，それは運用上の問題についての詳細な検討を促進しています．

　検討結果はすでに発表されています．望ましくない直流循環潮流を避けるという点と系統内の最適スラックノードの位置を見つけるという点の両方において，制御システムの広範囲にわたる協調が不可欠であることを示しています．

〈補足〉　"offshore"は「洋上の」，"feasibility study"は「実現可能性の検討」，"transmission system operator (TSO)"は「送電系統運用者」，"wind power plant"「風力発電所」，"market-driven"は「市場主導の」，"voltage source converter (VSC)"は「電圧形変換器(VSC)」，"high voltage direct current (HVdc)"は「直流高電圧」，"slack node"は「スラックノード」(電力系統の有効・無効電力を調整する発電機が接続されるノード)です．

　A voltage sag is a short-duration (up to a few seconds) reduction in voltage magnitude. The voltage temporarily drops to a lower value, e.g., from 230 V down to 170 V (74 % of nominal), and comes back again after approximately 150 ms. Despite their short duration, such events can cause serious problems for a wide range of equipment. Process-control equipment, computers, and adjustable-speed system drives are most notorious for their sensitivity to these sags (M. H. J. Bollen, p. 8, ©2001 IEEE. IEEE による許諾を得て転載).

電圧低下とは，短時間の(数秒までの)電圧振幅の低下のことです．たとえば 230 V から 170 V (公称値の 74 %)に電圧が一時的に低下し，その後約 150 ms でもとの電圧に戻ります．短時間にもかかわらず，このような事象は幅広い機器に深刻な問題を引き起こす可能性があります．プロセス制御装置，コンピュータ，可変速度システム駆動装置は，それらに影響を受けやすいことで知られています．

〈補足〉 "voltage sag"「電圧低下」，"process-control equipment"は「プロセス制御装置」，"adjustable-speed system drive"は「可変速度システム駆動装置」です．

3.3.2 高電圧現象に関する表現
Expressions for High-voltage Phenomena

ここでは，高電圧送電線および雷放電に関する例文を示します．

Ultrahigh-voltage (above 765-kV) power transmission research has been proceeding in the United States for a number of years. Since this type of research deals with the transport of very large quantities of power on big overhead lines at the very time when there are strong pressures to put the power underground for environmental reasons, the research effort may seem a little incongruous. However, it is precisely for environmental reasons that much of this work is of interest. The demand for power in the United States keeps increasing, year by year. One industry forecast suggests a summer peak of 453 GW by 1977, almost twice the 1969 level. The electrical energy output in 1977 is prediced to be over 2 500 TWh—an increase of 1 100 TWh in the 8 years beyond 1969. Even if these forecasts go far beyond the mark, there is still a great deal of new power to be supplied and new transmission capability must be built to carry it.

In the increased concerns of the present day for environment and pollution, one of the major reasons for burying this new transmission is to avoid visual pollution. However, the act of burying brings other pollution and environmental effects that may be even worse. One is a squandering of our natural resourses. It would take a massive amount of material in the form of expensive cables, shunt reactors, and other equipment and facilities to accomplish this. This material must be taken from our dwindling natural resources, melted and formed from the burning of our coal and other fuels (which add their own pollution). The inefficiencies of underground transmission eventually reflect adversely into our environment in many ways, some of which may be much worse than visual pollution.

On the other hand, UHV overhead transmission makes the most efficient means now available for the transmission of large blocks of power. It uses the least of the nation's raw materials, creates the least heat, requires the least amount of fuel to be continuously burned to supply its losses, and eliminates the necessity for much lower voltage line construction. One of the objectives of UHV overhead transmission research is to develop the technology of designing these major carriers of power so that they have a minimum impact on the ecology and the environment, yet can function effectively to supply the increasing power demands of the United States (Anderson et al., p. 1548, ©1971 IEEE. IEEEによる許諾を得て転載).

　超々高電圧(765 kV 以上)電力伝送の研究は，何年もの間，アメリカで進められてきました．この種の研究は，環境上の理由から電力伝送設備を地下に配置すべきであるという強い圧力があるまさにそのときに，巨大な架空線による大電力輸送を扱っているため，この研究に取り組むことは不適当に見えるかもしれません．しかし，この研究の大部分が興味深いのはまさに環境上の理由なのです．アメリカにおける電力需要は年々増え続けています．ある産業予測は，1977年までに夏のピークは453 GWになるであろうと示唆しています．これは1969年レベルのほぼ2倍です．1977年の電気エネルギー出力は2500テラワット時を超えると予想されています(1969年からの8年で1100テラワット時の増加)．たとえ，これらの予想が見当違いであったとしても，依然として多量の新電力を供給しなければなりませんし，それを輸送するための新しい送電設備を建設しなければなりません．

　今日，環境と汚染に対する関心が高まっているなかで，この新しい送電線を埋設しようとする主要な理由の一つは視覚公害を避けることです．しかし，埋設すると，さらに悪いかもしれないほかの公害や環境上の影響をもたらします．一つは，私たちの天然資源を浪費してしまうことです．これを成し遂げるためには，高価なケーブル，分路リアクトル，そ

のほかの機器や設備という形態で多量の原料が必要になります．これらの原料は，私たちの減少しつつある天然資源から採ってこなければなりませんし，石炭やほかの燃料を燃焼させることにより（これらも汚染を増やします），溶かされ，そして成型されなければなりません．地中送電の効率の悪さは，結局は，いろいろな形で私たちの環境に跳ね返ってきます．それらのなかのいくつかは視覚公害よりもはるかに悪いかもしれません．

　一方で，UHV 架空送電は，多量の電力を輸送するために今日利用できる最も効率のよい手段です．それは，この国の原料を最小限しか使用せず，最小限の熱しか発生せず，損失分を供給するために継続して燃やさなければならない燃料も最小限しか必要とせず，多くの低電圧送電線の建設の必要性をなくします．UHV 架空送電研究の目的の一つは，生態や環境に与える影響が最小になるようにしつつ，それでもアメリカの増えていく電力需要に応じるのに有効に機能できるように，これらの主要電力輸送手段の設計技術を開発することです．

〈補足〉 "ultrahigh-voltage (UHV) power transmission" は「超々高電圧(UHV)電力伝送」，"incongruous" は「不適当な」，"beyond the mark" は「見当違いで」，"visual pollution" は「視覚公害」，"squander" は「浪費する」，"shunt reactor" は「分路リアクトル」，"dwindling" は「減少しつつある」，"overhead transmission" は「架空送電」，"ecology" は「生態」です．

　Lightning is a transient, high-current electric discharge whose path length is measured in kilometers. The most common source of lightning is the electric charge separated in ordinary thunderclouds (cumulonimbus) although other types of clouds, such as nimbostratus, may occationally produce natural lightning. Well over half of all lightning discharges occur wholly within the cloud and are called intracloud discharges. The common cloud-to-ground lightning which lowers negative charge to Earth has been studied more extensively than other forms of lightning because of its practical interest (e.g., as the cause of injuries and death, disturbances in power and communication systems, and the ignition of forest fires) and because lightning below cloud level is more easily photographed and studied with optical instruments. Cloud-to-cloud and cloud-to-air discharges occur less frequently than either intracloud or cloud-to-ground lightning. All discharges other than cloud-to-ground are often lumped together and called cloud discharges (Uman, p. 1549, ©1988 IEEE. IEEEによる許諾を得て転載).

　雷は，長さ数kmの過渡的な大電流放電です．乱層雲のようなほかの種類の雲もときには自然雷を起こしますが，雷の最も一般的な源は通常の雷雲（積乱雲）内で分離された電荷です．すべての雷放電の優に半分以上は完全に雲内で発生し，雲内放電と呼ばれています．実際的な関心（たとえば，負傷や死，電力システムや通信システムにおける障害，そして森林火災発生の原因として）から，そして雲下の雷はより簡単に写真をとることができ，光学機器により研究しやすいため，負電荷を大地に下ろす通常の対地雷は，ほかの形態の雷より広く研究されてきました．雲間放電や大気放電は，雲内放電や対地雷に比べると，あまり発生しません．対地放電以外のすべての放電はしばしば一まとめにして扱われ，雲放電と呼ばれています．

〈補足〉　"lightning"は「雷」，"high current"は「大電流」，"electric(al) discharge"は「放電」，"cumulonimbus"は「積乱雲」，"nimbostratus"は「乱層雲」，"intracloud discharge"は「雲内放電」，"cloud-to-ground lightning"は「対地雷」，"cloud-to-cloud discharge"は「雲間放電」，"cloud-to-air discharge"は「大気放電」，"cloud discharge"は「雲放電」です．

3.3.3　エネルギー貯蔵に関する表現
Expressions for Storage of Energy

ここでは，超伝導磁気エネルギー貯蔵と電池に関する例文を示します．

　The variations in electric power usage have required the utilities and the manufacturers of power generating equipment to design power generating units capable of cycling and to develop methods of storing energy to meet varying power demands. Most of these demands are periodic, but the cycle time may vary from a few seconds to a year. The annual variation is usually accommodated by scheduling power equipment outage and major maintenance for low-demand seasons. The daily and weekly variations, however, are the most important because of the sheer magnitude of the power variations that may occur during periods as short as an hour. For example, the change in power demand is often from 60 percent of the peak load at 7 AM to 90 percent at 9 AM. On a system with a 2 000-MW peak load this variation is about 600 MW. Large coal-fired and nuclear power plants, which are the most economic generating units, are normally designed to operate at full or nearly full capacity with little or no power variation. Their expected life is considerably decreased when forced to cycle by large fractions of total output capacity. Gas turbines, old and intermediate-sized power plants, and energy storage units, which are designed for cycling or are less affected by changing power levels, are cycled to meet the daily and weekly power variations.

 A variety of power generating and energy storage technologies can satisfy the cyclic power demand. Some, such as gas turbines, hydroelectric, and pumped hydroelectric, have been used widely. Several new technologies, including compressed air, underground pumped hydro, batteries, and superconducting magnets show promise for possible future applications, and some have already seen limited applications though they are all still in the development process. With adequate development, and assuming reasonable costs can be achieved for these technologies, utilities in the future will be able to select the type of plant that will optimize their power generation capability in terms of cost and performance.

 Superconducting magnetic energy storage (SMES) is inherently very efficient and has siting requirements that are different from other technologies. Because of these characteristics, SMES has the potential of finding application in systems with large energy storage requirements. This recently conceived technology meets many of the utilities' requirements for diurnal storage. An unusual feature of SMES is the cost scaling with size, which is different from that for other storage devices. For a given design, the cost of a SMES unit is roughly proportional to its surface area and the required quantity of superconductor. The cost per unit of stored energy (megajoule or kilowatt-hour) decreases as storage capacity increases.

 In addition, the charge and discharge of a SMES unit is through the same device, a multiphase converter or Graetz bridge, which allows the SMES system to respond within tens of milliseconds to power demands that could include a change from maximum charge rate to maximum discharge power. This rapid response allows a diurnal storage unit to provide spinning reserve and to improve system stability. Both the converter and the energy storage in the coil are highly efficient as there is no conversion of energy from one form to another as in pumped hydro, for example, where the electrical energy is converted to mechanical energy and then back again. The major loss during storage is the energy required to operate the refrigerator that maintains the superconducting coil at 1.8 K. SMES, because of these characteristics and because it can be easily sited, has the potential of finding extensive application on electric utility systems (Hassenzahl, p. 1089, ©1983 IEEE. IEEE による許諾を得て転載).

　電力消費は変動するため，電力会社や発電機メーカは，オンオフを繰返しできる発電機を設計し，そして変化する電力需要に対応できるエネルギー貯蔵法を開発する必要がありました．これらの需要のほとんどは周期的ですが，その周期は数秒から1年までさまざまです．電力需要の年変化は，通常は，低需要期間における電力機器の停止や大規模保守整備の計画を組むことにより調整されます．しかし，1時間程度の短時間に生じる電力変動は大きいため，日変化や週変化は最も重要です．たとえば，電力需要の変化は，午前7時にピーク負荷の60％であったのが，午前9時には90％になることも頻繁にあります．ピーク負荷が2 000 MWの系統の場合，この変化は約600 MWになります．大規模石炭火力発電所と原子力発電所は最も経済性が高く，通常，それらはフル稼働か，非常にわずかな電力変動を伴うか，あるいは変動なしで，ほぼフル稼働するように設計されています．それらに全出力容量のかなりの割合の変化を強いると，それらの予想耐用年数は大幅に短くなります．ガスタービン，旧型や中規模発電所およびエネルギー貯蔵装置はオンオフを繰返しできるように設計されているか，あるいは電力レベルの変動にあまり影響を受けません．これらは電力の日変化や週変化に対応するためオンオフされます．

　さまざまな発電技術およびエネルギー貯蔵技術が周期的な電力変動に対応することができます．それらのうちのいくつか，たとえば，ガスタービン，水力発電および揚水式水力発電が広く用いられてきました．圧縮空気，地下揚水式発電，電池および超伝導磁石を含むいくつかの新技術は，将来，適用される見込みがあります．今も開発過程にありますが，いくつかはすでにかぎられた範囲において適用されています．これらの

3.3 電力・エネルギー分野における表現

技術が十分に開発され，そして妥当なコストが実現されれば，将来，電力会社は，コストと性能の点から発電力を最適化する発電所形態を選ぶことができるでしょう．

超伝導磁気エネルギー貯蔵(SMES)は本質的に高効率で，ほかの技術とは異なる(有利な)据付条件をもっています．これらの特徴のため，SMESは大エネルギーの貯蔵を必要とする系統に適用される可能性があります．この最近考え出された技術は，電力会社が日変化用の電力貯蔵に対して要求する条件の多くを満たしています．SMESのほかとは異なる特徴は大きさに対するコストで，ほかの貯蔵装置のそれとは異なっています．ある設計に対して，SMES装置のコストはその表面積と必要とされる超伝導体の量におおむね比例します．単位貯蔵エネルギー(メガジュールあるいはキロワット時)あたりのコストは，貯蔵容量が増えるほど下ります．

そのうえ，SMES装置の充放電は，多相コンバータあるいはグレーツブリッジと呼ばれる同一の機器を介して行われます．これにより，SMESシステムは，その最大充電量から最大放電量への変化を含む電力需要に対しても数十ミリ秒以内に応答することができます．この早い応答により，日変化用貯蔵装置は瞬時予備力を提供でき，そして系統安定度を改善することができます．たとえば，電気エネルギーから力学的エネルギーに変換し，そして再び電気エネルギーに戻す揚水発電のようにエネルギー形態を変換する必要がないため，コンバータおよびコイル内のエネルギー貯蔵はともに非常に高効率です．貯蔵中の主な損失は，超伝導コイルを1.8ケルビンに保つ冷却装置の運転に要するエネルギーです．SMESは，これらの特長をもち，そして据付が容易であることから，電気利用システムの広範囲に適用される可能性があります．

〈補足〉 "utility"は「(電気，ガス，水道などの)公益事業者」，"manufacturer"は「メーカ(製造業者)」，"periodic"は「周期的な」，"annual"は「1年の」，"outage"は「停電(停止)」，"coal-fired (power) plant"は「石炭火力発電所」，"nuclear power plant"は「原子力発電所」，"gas turbine"は「ガスタービン」，"energy storage unit"は「エネルギー貯蔵装置」，"pumped hydroelectric"は「揚水式水力発電の」，"compressed air"は「圧縮空気」，"battery"は「電池」，"superconducting magnet"は「超伝導磁石」，"superconducting magnetic energy storage (SMES)"は「超伝導磁気エネルギー貯蔵」，"site"は「立地を決める(配置する)」，"diurnal"は「日ごとの(日中の)」，"be proportional to..."は「...に比例する」，"spinning reserve"は「瞬時予備力」，

第 3 章　電気電子工学分野における諸表現

"refrigerator" は「冷却装置」です．

> The greatest effort to demonstrate the use of large batteries as a tool to manage power demand in a utility grid began in the 1990s in Japan with the development of the Sodium Sulfer (NaS) battery system capable of delivering at least six hours of battery runtime on a daily basis. This concept was first demonstrated in the United States in 2006 with a 1.0-MW NaS battery deployed in a utility substation to deal with summer peak loads on a 20 MVA station transformer.
>
> By 2010 NaS battery installations totalled 365 MW-s worldwide with 300 MW-s in service in Japan. In 2009 the world's largest battery (34 MW-s) went into service in support of a wind farm in Northern Japan (Roberts, and Sandberg, p. 1141, ©2011 IEEE. IEEE による許諾を得て転載).
>
> 　電力系統における電力需要に対処するための一手段として大形電池を利用することを実証するための最大の取組みが，少なくとも毎日 6 時間の電池駆動時間は電力供給できるナトリウム硫黄(NaS)電池システムの開発とともに，1990 年代に日本で始まりました．この構想は，20 メガボルトアンペアの変圧器において夏のピーク負荷に対処するために変電所に置かれた 1.0 メガワットの NaS 電池を用いて，2006 年にアメリカで初めて実証されました．
>
> 　2010 年までの NaS 電池の全世界における設置総容量は 365 メガワット秒，日本で稼働中のものが 300 メガワット秒です．2009 年に世界最大の電池(34 メガワット秒)が，北日本にある風力発電所をサポートするために稼働し始めました．

〈補足〉　"sodium sulfer (NaS) battery" は「ナトリウム硫黄(NaS)電池」，"battery runtime" は「電池駆動時間」，"wind farm" は「風力発電所」です．

> By the end of 2022 about 9 GW of energy storage had been added to the U.S. grid since 2010, adding to the roughly 23 GW of pumped storage hydropower (PSH) installed before that. Of the new storage capacity, more than 90% has a duration of 4 hours or less, and in the last few years, Li-ion batteries have provided about 99% of new capacity (Denholm et al., p. vi, National Renewable Energy Laboratory, 2023).

2010年以降2022年末までに，約9GWのエネルギー貯蔵設備が米国の電力系統に追加され，それ以前に設置された約23GWの揚水発電設備を増強しました．新規追加されたエネルギー貯蔵容量のうちの90%以上は4時間以下の持続時間です．ここ数年においては，リチウムイオン電池が新規追加容量の約99%を供給しています．

〈補足〉 "pumped storage hydropower"は「揚水発電」，"Li-ion (lithium-ion) battery"は「リチウムイオン電池」です．

3.4 電子デバイスに関する表現
Expressions for Electronics Devices

本節では，電子デバイス分野で用いられる諸表現を示します．

3.4.1 ダイオードとトランジスタに関する表現
Expressions for Diodes and Transitors

ここでは，シリコン，ゲルマニウム，n形およびp形半導体，ダイオード，バイポーラ接合トランジスタに関する例文を示します．

The most important semiconductors used in the electronics industry are silicon and germanium. As the temperature of these materials is raised above room temperature, the resistivity is reduced and ultimately a point is reached where they effectively become conductors. For this reason, silicon should not operate at a working temperature in excess of 150 ℃ to 200 ℃, depending on its purity, and germanium should not operate at a working temperature in excess of 75 ℃ to 90 ℃, depending on its purity. As the temperature of a semiconductor is reduced below normal room temperature, the resistivity increases until, at very low temperatures the semiconductor becomes an insulator (J. Bird, Electrical Circuit Theory and Technology, p. 138, ©2003 Newnes. Taylor & Francis Books UKによる許諾を得て転載).

エレクトロニクス産業において用いられている最も重要な半導体はシリコンとゲルマニウムです．これらの物質の温度が室温より上げられる

> と，それらの抵抗率が低下し，最終的には，それらが事実上導体になる点にいたります．この理由で，シリコンは，純度に応じて150℃から200℃を超える動作温度で使用するべきではなく，またゲルマニウムは，純度に応じて75℃から90℃を超える動作温度で使用するべきではありません．半導体の温度が通常の室温より低くなると，その抵抗率が上昇し，非常に低い温度においては，半導体は絶縁体になります．

〈補足〉 "silicon"は「シリコン」，"germanium"は「ゲルマニウム」，"room temperature"は「室温」，"purity"は「純度」，"in excess of ..."は「... を超えて」，"insulator"は「絶縁体」です．

> Adding extremely small amounts of impurities to pure semiconductors in a controlled manner is called doping. Antimony, arsenic and phosphorus are called n-type impurities and form an n-type material when any of these impurities are added to silicon or germanium. The amount of impurity added usually varies from 1 part impurity in 10^5 parts semiconductor material to 1 part impurity to 10^8 parts semiconductor material, depending on the resistivity required. Indium, alminium and boron are called p-type impurities and form a p-type material when any of these impurities are added to a semiconductor.
> 　In semiconducting materials, there are very few charge carriers per unit volume free to conduct. This is because the "four electron structure" in the outer shell of the atoms (called valency electrons), form strong covalent bonds with neighboring atoms, resulting in a tetrahedral structure with the electrons held fairly rigidly in place.
> 　Antimony, arsenic and phosphorus have five valency electrons and when a semiconductor is doped with one of these substances, some impurity atoms are incorporated in the tetrahedral structure. The "fifth" valency electron is not rigidly bonded and is free to conduct, the impurity atom donating a charge carrier.
> 　Indium, aluminium and boron have three valency electrons and when a semiconductor is doped with one of these substances, some of the semiconductor atoms are replaced by impurity atoms. One of the four bonds associated with the semiconductor material is deficient by one electron and this deficiency is called a hole. Holes give rise to conduction when a potential difference exists across the semiconductor material

due to movement of electrons from one hole to another. The resulting material is p-type material containing holes (J. Bird, Electrical Circuit Theory and Technology, pp. 138-139, ⓒ2003 Newnes. Taylor & Francis Books UK による許諾を得て転載).

　真性半導体に非常に少量の不純物を制御しながら加えることはドーピングと呼ばれています．アンチモン，ヒ素およびリンは n 形不純物と呼ばれており，これらの不純物のうちのどれかがシリコンかゲルマニウムに加えられると n 形物質を形成します．加えられる不純物の量は，通常，必要とする抵抗率に依存して，半導体物質 10^5 片に不純物 1 片から半導体物質 10^8 片に不純物 1 片の範囲で変化します．インジウム，アルミニウムおよびホウ素は p 形不純物と呼ばれており，これらの不純物のうちのどれかが半導体に加えられると p 形物質を形成します．

　半導体物質中には，自由に伝導できる単位体積あたりの電荷担体はほとんどありません．これは，（価電子と呼ばれる）原子の外殻における「4 電子構造」が隣接する原子と強力な共有結合を形成しており，非常に堅く拘束されたそれらの電子とともに 4 面体構造となるからです．

　アンチモン，ヒ素およびリンは五つの価電子をもっています．これらの物質の一つが半導体に添加されると，不純物原子のいくつかは 4 面体構造に組み込まれます．「5 番目の」価電子は堅くは結合されず，自由に伝導し，不純物原子は電荷担体を提供します．

　インジウム，アルミニウムおよびホウ素は三つの価電子をもっています．これらの物質の一つが半導体に添加されると，半導体の原子のいくつかは不純物原子に置き換えられます．半導体物質の四つの結合のうちの一つが電子 1 個を欠きます．この欠陥は正孔と呼ばれています．半導体物質に電位差が存在する場合には，ある正孔から別の正孔への電子の移動により，正孔は伝導を生じさせます．この結果として生じる物質が正孔を含む p 形物質です．

〈補足〉 "impurity" は「不純物」，"pure（または intrinsic）semiconductor" は「真性半導体」，"doping" は「ドーピング（不純物添加）」，"antimony" は「アンチモン」，"arsenic" は「ヒ素」，"phosphorus" は「リン」，"indium" は「インジウム」，"boron" は「ホウ素」，"charge carrier" は「電荷担体」，"valency electron" は「価電子」，"covalent bond" は「共有結合」，"tetrahedral" は「4 面体の」，"donate" は「提供する」，"deficient" は「不完全な（欠けた）」，"deficiency" は「欠陥（不足）」，"hole" は「正孔（ホール）」，"give rise to ..." は「...

を生じさせる」です．

> When the positive terminal of a battery is connected to the p-type material of a silicon diode and the negative terminal to the n-type material, the diode is forward biased. Because of like charges repelling, the holes in the p-type material drift towards the junction. Similarly the electrons in the n-type material are repelled by the negative bias voltage and also drift towards the junction. The width of the depletion layer and size of the contact potential are reduced. For applied voltages from 0 to about 0.6 V, very little current flows. At about 0.6 V, majority carriers begin to cross the junction in large numbers and current starts to flow. As the applied voltage is raised above 0.6 V, the current increases exponentially.
>
> When the negative terminal of the battery is connected to the p-type material and the positive terminal to the n-type material, the diode is reverse biased. The holes in the p-type material are attracted towards the negative terminal and the electrons in the n-type material are attracted towards the positive terminal. This drift increases the magnitude of both the contact potential and the thickness of the depletion layer, so that only very few majority carriers have sufficient energy to surmount the junction.
>
> The thermally excited minority carriers, however, can cross the junction since it is, in effect, forward biased for these carriers. The movement of minority carriers results in a small constant current flow. As the magnitude of the reverse voltage is increased a point will be reached where a large current suddenly starts to flow. The voltage at which this occurs is called the breakdown voltage. This current is due to two effects:
>
> (i) the Zener effect, resulting from the applied voltage being sufficient to break some of the covalent bonds, and
>
> (ii) the avalanche effect, resulting from the charge carriers moving at sufficient speed to break covalent bonds by collision.
>
> A Zener diode is used for voltage reference purposes or for voltage stabilization (J. Bird, Electrical Circuit Theory and Technology, pp. 142-143, ©2003 Newnes. Taylor & Francis Books UKによる許諾を得て転載).

電池の正極がシリコンダイオードのp形物質に，負極がn形物質に接続されると，そのダイオードは順方向にバイアスされます．同種の電荷の反発により，p形物質の正孔は(pn)接合部に向かってドリフトします．同様に，n形物質の電子は負のバイアス電圧により反発を受け，接合部に向かってドリフトします．空乏層の幅と接触電位の大きさは小さくなります．印加電圧が0から約0.6Vの場合には，ほとんど電流は流れません．約0.6Vのときに，多数キャリヤが大量に接合部を横切り始め，電流が流れ始めます．印加電圧が約0.6Vを超えると，電流は指数関数的に増加します．

　電池の負極がp形物質に，正極がn形物質に接続されると，そのダイオードは逆方向にバイアスされます．p形物質の正孔は負極に引き付けられ，n形物質の電子は正極に引き付けられます．このドリフトは，接触電位および空乏層の厚みの両方を大きくします．その結果，多数キャリヤのほとんどは，接合部を突破できるのに十分なエネルギーをもちません．

　しかし，熱的に励起された少数キャリヤにとっては，事実上順方向バイアスであるため，それらは接合部を横切ることができます．少数キャリヤの移動は，小さな一定電流となります．逆方向電圧が高められると，大電流が突然流れ始める点にいたります．これが生じる電圧は降伏電圧と呼ばれています．この電流は，二つの効果に起因しています．

(i) 共有結合のいくつかを破壊するのに十分な印加電圧により生じるツェナー効果と，

(ii) 衝突により共有結合のいくつかを破壊するのに十分な速度で移動する電荷担体により生じるなだれ効果です．

　ツェナーダイオードは，電圧基準あるいは電圧安定化に使われています．

〈補足〉 "forward biased"は「順方向にバイアスされた」，"repelling"は「反発」，"drift"は「ドリフトする」，"depletion layer"は「空乏層」，"majority carrier"は「多数キャリヤ」，"reverse biased"は「逆方向にバイアスされた」，"surmount"は「突破する」，"thermally excited"は「熱的に励起された」，"minority carrier"は「少数キャリヤ」，"in effect"は「事実上」，"breakdown voltage"は「降伏電圧(絶縁破壊電圧)」，"Zener effect"は「ツェナー効果」，"avalanche effect"は「なだれ効果」，"voltage stabilization"は「電圧安定化」です．

The bipolar junction transistor consists of three regions of semiconductor material. One type is called a p-n-p transistor, in which two regions of p-type material sandwich a very thin layer of n-type material. A second type is called an n-p-n transitor, in which two regions of n-type material sandwich a very thin layer of p-type material. Both of these types of transistors consist of two p-n junctions placed very close to one another in a back-to-back arrangement on a single piece of semiconductor material.

The two p-type material regions of the p-n-p transistor are called the emitter and collector and the n-type material is called the base. Similarly, the two n-type material regions of the n-p-n transistor are called the emitter and collector and the p-type material region is called the base.

Transistors have three connecting leads and in operation an electrical input to one pair of connections, say the emitter and base connections can control the output from another pair, say the collector and emitter connections. This type of operation is achieved by appropriately biasing the two internal p-n junctions. The base-emitter junction is forward biased and the base-collector junction is reverse biased (J. Bird, Electrical Circuit Theory and Technology, p. 145, ©2003 Newnes. Taylor & Francis Books UK による許諾を得て転載).

　バイポーラ接合トランジスタは半導体物質の三つの領域からなります．第一のタイプは p-n-p トランジスタと呼ばれており，p 形物質の 2 領域が n 形物質の非常に薄い層を挟んでいます．第二のタイプは n-p-n トランジスタと呼ばれており，n 形物質の 2 領域が p 形物質の非常に薄い層を挟んでいます．これらのトランジスタは両方とも，一体成形の半導体物質に背中合せで互いに非常に近接して配置された二つの p-n 接合からなります．

　p-n-p トランジスタの二つの p 形領域はエミッタとコレクタと呼ばれており，n 形領域はベースと呼ばれています．同様に，n-p-n トランジスタの二つの n 形領域はエミッタとコレクタと呼ばれており，p 形領域はベースと呼ばれています．

　トランジスタは三つの接続リード線をもっています．動作中は，1 対の接続部，たとえばエミッタとベースの接続部への電気的入力が別対の接続部，たとえばコレクタとエミッタの接続部からの出力を制御できま

す．この種の動作は二つの内部 p-n 接合部に適切なバイアスをかけることで実現されます．ベースとエミッタの接合部は順方向にバイアスをかけられ，ベースとコレクタの接合部は逆方向にバイアスをかけられます．

〈補足〉 "bipolar junction transistor" は「バイポーラ接合トランジスタ」，"back-to-back" は「背中合せの」，"connecting lead" は「接続リード線」，"say" は「たとえば」です．

3.4.2 集積回路に関する表現
Expressions for Large-scale Integrated Circuits

ここでは，集積回路とマイクロプロセッサに関する例文を示します．

Although Intel began as a memory chip company, in 1969 we took on a project for Busicom of Japan to design eight custom large scale integrated (LSI) chips for a desktop calculator. Each custom chip had a specialized function—keyboard, printer, display, serial arithmetic, control, etc. With only two designers, Intel did not have the manpower to do that many custom chips. We needed to solve their problem with fewer chip designs. Ted Hoff chose a programmed computer solution using one complex logic chip (CPU) and two memory chips; memory chips are repetitive and easier to design. Intel was a memory chip company, so we found a way to solve our problem using memory chips!

In 1970 Intel designers implemented a 4-b computer on three LSI chips (CPU, ROM, RAM) housed in 16-pin packages. Reducing the data word to 4-b (for a BCD digit) was a compromise between 1-b serial calculator chips and conventional 16-b computers. The scaled down 4-b word size made the CPU chip size practical (~2 200 transistors). We used the 16-pin package, because it was the only one available in our company. This limited pin count forced us to time multiplex a 4-b bus. This small bus simplified the printed circuit board (PCB), as it used fewer connections. However, the multiplexing logic increased chip area of the specialized ROM/RAM memory chips, which then had to have built-in address registers. Increasing the transitor count to save chip connections was a novel idea. In school we learned to minimize logic, not interconnections! Later, LSI "philosophers" would preach "logic is free"

第3章　電気電子工学分野における諸表現

(Mazor, pp. 1601-1602, ©1995 IEEE. IEEEによる許諾を得て転載).

　インテル社はメモリ素子の会社として始まりましたが，1969年に日本のビジコン社の電卓用の八つのカスタム大規模集積回路（LSI）素子を設計する事業を請け負いました．カスタム素子はそれぞれ特殊化された機能（キーボード，プリンタ，ディスプレイ，直列演算，制御など）をもっていました．インテル社は，たった2名の設計者のみで，これほど多くのカスタム素子を設計するマンパワーをもっていませんでした．私たちは，より少ない数の素子設計で，これらの問題を解決する必要がありました．テッド・ホフは，複合論理素子（CPU）一つとメモリ素子二つを用いたプログラム化したコンピュータ解法を選びました．メモリ素子は反復構造で，設計がより簡単なのです．インテル社はメモリ素子の会社であったので，私たちは，私たち自身の問題をメモリ素子で解決する方法を見つけたのです！

　1970年に，インテル社の設計者たちは，16ピンのパッケージに格納した三つのLSI素子（CPU，ROM，RAM）上で4ビットの電子計算機を実現しました．4ビット（2進化10進数のため）へのワードデータの縮小は，1ビットの直列計算機素子と従来の16ビット電子計算機の間での妥協によるものでした．この縮小した4ビットワードの大きさは，CPU素子を実用的な大きさにしました（トランジスタ数2200まで）．16ピンのパッケージを用いたのは，それが私たちの会社で唯一手に入るものであったからです．このかぎられたピン数は，私たちに，4ビットバスの時分割多重化を強いりました．この小さなバスは，接続点が少なくなるため，プリント回路基板（PCB）を単純化しました．しかし，多重化論理は，特殊化されたROM/RAMメモリ素子領域を大きくしました．それらは，その後，内蔵アドレスレジスタをもたなければならなくなりました．素子の接続数を抑えるためにトランジスタ数を増やすことは斬新なアイデアでした．学校で，私たちは，接続数ではなく論理を最小化することを学びました！　後になって，LSI「哲学者たち」は「論理は無料」と説いたのでした．

〈補足〉　"memory chip"は「メモリ素子」，"large scale integrated (LSI) chips"は「大規模集積回路（LSI）素子」，"desktop calculator"は「卓上電子計算機（電卓）」，"serial arithmetic"は「直列演算」，"logic chip"は「論理素子」，"read-only memory (ROM)"は「読出し専用メモリ」，"random-access memory (RAM)"は「随時アクセスメモリ」，"house"は「格納する」，"binary-coded decimal (BCD)"は「2進化10進数」，"multiplex"は「多重化する」，"printed circuit board (PCB)"は「プリント回路基板」，"built-in"は「内蔵の（組込みの）」，"address register"は「アドレスレジスタ」です．

> In 1987, Intel's W. Davidow, vice president of the microcomputer group, rushed to staff a 16-b microcomputer development project. It was to have around 30 000 transistors, 12 times more than the 4004. This new computer had multiplication and division and a host of other new features. However, it was constrained to be upwardly compatible with the 8080 (and 8008). Accordingly, the designers decided to keep the 16-b basic addresses and to use segment registers to get extended 20-b addresses. Two versions were created—the 8088 had an 8-b data bus for compatibility with 8-b memory systems, and the 16-b 8086. With 1-megabyte of memory addressing, this processor was a serious contender in the computer market place. This chip density required to match the 16-b minicomputers was "arriving" as had been predicted.
>
> The decision by IBM to use the 8088 in a word processor and personal computer created enormous market momentum for Intel. The 186, 286, 386, 486 followed over the next 15 years, with some shadow of 8008 features still apparent. These components would be "truly pervasive" (Mazor, pp. 1604-1605, ©1995 IEEE. IEEEによる許諾を得て転載).
>
> 　1987年に，インテル社のマイクロコンピュータグループの統括責任者であったW．ダビドウは，16ビットのマイクロコンピュータ開発計画へのスタッフの配置を急ぎました．それは，4004の12倍にもなる約30 000個のトランジスタを搭載することになっていました．この新しいコンピュータは乗算，除算，およびほかにも多くの新しい特長をもっていました．しかし，それは8080(そして8008)と上位互換性をもつことを強いられていました．それゆえ，設計者たちは，16ビットの基本アドレスを保ち，セグメントレジスタを20ビットに拡張することを決めました．二つのバージョンが作られました(8ビットメモリシステムとの互換性のための8ビットデータバスをもつ8088と，16ビットの8086です)．1メガバイトのメモリアドレスをもったこのプロセッサは，コンピュータ市場において重要なものとなりました．16ビットのミニコンピュータと整合をとるために必要とされたこの素子密度に，予期されたとおりにたどり着きました．

第 3 章　電気電子工学分野における諸表現

> 　8088 をワードプロセッサとパーソナルコンピュータに使うという IBM 社の決定は，インテル社への大きな市場の勢いを作り出しました．その後の 15 年間に 186，286，386，486 が続きました．それらには，8008 の特長の名残が今も明らかです．これらの素子は本当に普及していました．

〈補足〉"a host of" は「多くの」，"upwardly compatible with…" は「…と上位互換性のある」です．

3.4.3　光電子デバイスに関する表現
Expressions for Optoelectronics Devices

ここでは，発光ダイオードに関する例文を示します．これは，毎月約 25 万部も発行されている IEEE Spectrum という雑誌に掲載された記事です（3.5.3 の例文も IEEE Spectrum からの引用です）．IEEE Transactions や IEEJ Transactions のような論文雑誌とは異なり，このような雑誌では，さまざまな専門分野の読者の興味を引くため，カジュアルでセンセーショナルな表現が多用された記事が掲載されています．

> 　Back in the 20th century, just about the only light-emitting diode (LED) you normally saw was the one that lit up when your stereo was on. By the noughties, tiny light-emitting diodes were also illuminating the display and keypads of your mobile phone. Now they are backlighting your notebook screen, and soon they will replace the incandescent and compact fluorescent lightbulbs in your home.
> 　This revolution in light comes from the ever-greater bang the LED delivers per buck. With every decade since 1970, when the red LEDs hit their stride, they have gotten 20 times as bright and 90 percent cheaper per watt; the relation is known as Heitz's Law, and it applies also to yellow and blue LEDs, which were commercialized much later. The forerunners of the white LEDs that are now going into lightbulbs were the chips that backlit handsets starting about a decade ago. Back then, they used tens of milliamps and consumed a watt for every 10 lumens of light they produced. They were also tiny—just 300 micrometers on a side. Since then, the chips have more than tripled in size, to a millimeter square or more, current has shot up to an ampere or so, and efficiency

has rocketed to around 100 lm/W. They now have everything they need to dominate lighting, except for a low enough price. That, however, will soon come, too.

Even now, white LEDs are competitive wherever replacing a burned-out lamp is inconvenient, such as in the high ceilings and twisty staircases of Buckingham Palace, because LEDs last 25 times as long as Edison's bulbs. They have a 150 percent edge in longevity over compact fluorescent lights (CFLs), and unlike CFLs, LEDs contain no toxic mercury. That means it is not a pain to dispose of them, and you do not have to worry that your house has become a hazard zone if one breaks (Ross, p. 38, ©2011 IEEE. IEEE による許諾を得て転載).

20 世紀には，普段見るほとんど唯一の発光ダイオード(LED) は，ステレオがついているときに点灯する発光ダイオードでした．21 世紀の最初の 10 年間には，非常に小さい発光ダイオードがディスプレイや携帯電話のキーパッドを明るくしていました．今では，ノートパソコンの画面を背面から照らし，まもなく，家庭の白熱電球や小形蛍光灯に取って代るでしょう．

この光の革命は，これからもますます高くなる LED のコストパフォーマンスによってもたらされています．LED が本領を発揮した 1970 年以来，10 年ごとに，明るさは 20 倍になり，1 ワットあたりの価格は 90％低くなってきました．この関係はハイツの法則として知られており，ずっと後になってから商品化された黄色および青色 LED にも適用できます．今や電球内に入りつつある白色 LED の先駆けは，10 年ほど前を発端とする電話機を背面から照らす素子でした．当時は，それらの素子は，数十ミリアンペアを使い，10 ルーメンの光あたり 1 ワットを消費していました．それらは，また，一辺 300 マイクロメートル程度と非常に小さいものでした．それ以来，LED 素子の大きさは 3 倍以上，1 ミリメートル四方あるいはそれ以上になり，電流は 1 アンペア程度に急上昇し，効率は 1 ワットあたり 100 ルーメンにも上昇しました．LED 素子は，今や，照明を独占するために必要な，十分安い価格以外のすべてをもっています．すぐにそうなるでしょう．

LED はエジソン電球よりも 25 倍長持ちしますので，バッキンガム宮殿の高い天井や曲がりくねった階段のように，切れた電球を取り換えるのが不便な場所では，LED は今でも競争力があります．LED は，小形蛍光灯に比べて，寿命で 150％優位にあります．小形蛍光灯と異なり，

第 3 章　電気電子工学分野における諸表現

> 有毒水銀も含んでいません．これにより，LED を捨てるのが苦痛ではなくなります．また，LED が割れて，あなたの家が危険地帯になってしまうことを心配する必要もありません．

〈補足〉　"light-emitting diode (LED)" は「発光ダイオード (LED)」，"lit" は "light"「発光する」の過去・過去分詞形，"noughties" は「21 世紀の最初の 10 年間」(低俗な意味を示すスラングと同じ発音ですので，論文には不向き)，"mobile phone" は「携帯電話」，"incandescent bulb" は「白熱電球」，"fluorescent lightbulb" は「蛍光灯」，"bang per buck" は「コストパフォーマンス」(論文には不向き．なお，俗語で buck は 1 ドルを表します)，"hit one's stride" は「本領を発揮する」(論文には不向き．ここでは，"achieved commercial success" を意味しています)，"forerunner" は「先駆け (先駆者)」，"handset" は「電話機」，"lumen" は「ルーメン (光束の単位)」，"shoot up to..." は「...に急上昇する」(論文には不向き)，"rocket to..." は「...に急上昇する」(論文には不向き)，"Edison's bulb" は「エジソン電球」(エジソン電球は炭化竹フィラメント電球を厳密には意味します)，"burned-out lamp (burnt-out lamp)" は「切れた電球」(バッキンガム宮殿に LED 照明が取り付けられたのは 2006 年)，"longevity" は「寿命」，"toxic mercury" は「有毒水銀」(LED 廃棄物が環境に全く無害であるともいい切れません) です．最後の 1 文はセンセーショナルすぎます．

発光ダイオードに関するこの例文を，ほかの関連する情報も含めて学術論文誌の記事のように書き換えたものを以下に示します．

> The use of light-emitting diodes (LEDs) in consumer devices has changed considerably since their initial development by Holonyak in 1962. After initial success in seven-segment LED displays for calculators, the technology was displaced by liquid crystal displays (LCDs) with lower power consumption in the late 1970s. The role of LEDs decreased in the 1990s. They were usually seen only as simple red power indicators. However, the light output and efficiency of LEDs continued to increase with time. There is now a renaissance of applications that include headlights for vehicles as well as backlights for notebook computers and the highest quality high definition television (HDTV) displays.
>
> In 2000, Heitz reviewed the light output of LEDs as a function of time and proposed a relation, similar to the more familiar "Moore's Law" for

semiconductors. Heitz suggested that, with each decade, the cost per lumen would fall by a factor of 10, and the amount of light generated per LED package would increase by a factor of 20, for each specific color technology. This relation has proved to be valid for the original red LEDs, as well as the yellow, orange, green and blue LEDs which were commercialized at later dates. In the period 2001–2011, white LEDs, combining yellow and blue wavelengths for illumination, tripled in size from 0.3 mm to 1 mm per side, with drive current of 1 A rather than 10 mA and output efficiency of 100 lumens per Watt (lm/W) compared to 10 lm/W.

It is anticipated that LEDs will displace both incandescent and compact fluorescent lamps (CFL) in household lighting applications in the decade 2010 to 2020. The LEDs retain some desirable features of incandescent bulbs, such as selectable color temperature and efficient modulation of intensity. They also offer the long service life and high efficiency of CFL. In addition, LED lights have improved ruggedness for mobile application. They also represent reduced environmental impact. The life-cycle cost of LED lighting does not involve any release of trace amounts of mercury. There is typically 3–5 mg inside a compact fluorescent bulb, and 6 mg will be released from coal-fired power plant emissions to provide the excess heat energy to an incandescent bulb of the same light intensity, based on the emission rate of 12 μg/kWh in the USA.

The US Department of Energy posted an L Prize® of $US 10 Million to promote the development of replacement LED bulbs. This prize was awarded for the 60-W category to Phillips on 3 August 2011. The winning LED bulb has a standard A19 form factor and uses less than 10 watts to produce a non-directional light output of 940 lumens. The bulb was released commercially in February 2012 at a cost of approximately $US 25. With the declining cost of the LEDs compared to their packaging and drive circuits, it will be interesting to follow the future correspondence to Heitz' Law.

1962年のホロニアックによるLEDの初期開発以来，消費者機器における発光ダイオード(LED)の利用はずいぶん変化してきました．電卓用の7セグメントLEDディスプレイにおいて初期に成功した後，1970

年代後半に低消費電力液晶ディスプレイ(LCD)により，その技術は取って代られました．1990年代には，LEDの役割は減少しました．それらは，簡易な赤色の電源表示灯としてのみ見られました．しかし，LEDの光出力と効率は時間とともに上昇し続けました．今は，ノートパソコンや高質の高精細度テレビ(HDTV)のバックライトや自動車のヘッドライトを含む応用の再興期です．

2000年に，ハイツはLEDの光出力を時間の関数として再調査し，より広く知られている半導体に関するムーアの法則に似た関係を提案しました．ハイツは，10年ごとに，1ルーメンあたりのコストは10分の1倍に低下し，1パッケージあたりのLEDの光出力は20倍に上昇することを，各色のLED技術に対して示唆しました．この関係は最初の赤色LEDに対してだけではなく，後に商品化された黄色，赤茶色，緑色および青色LEDに対してもなりたつことが証明されました．2001年から2011年の間に，黄色と青色波長を組み合せた白色LEDは，駆動電流が10ミリアンペアから1アンペアに，出力効率が1ワットあたり10ルーメンから100ルーメンになり，一辺あたりの大きさが0.3ミリメートルから，その3倍の1ミリメートルになりました．

2010年から2020年までの10年間に，LEDは家庭の照明における白熱電球と小形蛍光灯(CFL)の両方に取って代ることが予想されます．LEDは，色温度を選べることや強度を調整できることなど，白熱電球の望ましい特長を保持しています．また，小形蛍光灯の長寿命と高効率も備えています．さらに，LEDライトは携帯使用のための耐久性も改善します．また，環境への影響も小さくなります．LED照明の生涯コストは，微量の水銀の除去を全く含みません．小形蛍光灯内に通常は3から5ミリグラムの水銀が含まれています．また，アメリカにおける1キロワット時あたり12マイクログラムの(水銀)放出率に基づくと，同じ光出力の白熱電球に過度の熱エネルギーを供給するためには，石炭火力発電所から6ミリグラムの水銀が放出されることになります．

アメリカのエネルギー省は(白熱電球の)代替LED電球の開発を促進するため，賞金1 000万米ドルのL Prize®を公示しました．2011年8月3日に，この賞は，60ワット部門において，フィリップス社に授与されました．この賞を獲得したLED電球は，標準的なA19形で，940ルーメンの無指向性の光を出力するのに10ワットも使用しません．このLED電球は，2012年2月に約25米ドルで発売されました．パッケージングや駆動回路に比べて，ますます下がっていくLEDのコストとともに，ハイツの法則に将来も一致していくことを見守るのは興味深い

〈補足〉 "liquid crystal display (LCD)" は「液晶ディスプレイ」，"high definition television (HDTV)" は「高精細度テレビ」，"trace amounts" は「微量の」です．2023年の「水銀に関する水俣条約」第5回締約国会議で，一般照明用蛍光灯の製造と輸出入を2027年末までに廃止することが決定されています．

3.5 情報通信に関する表現
Expressions for Information and Communications

本節では，情報通信分野で用いられる諸表現を示します．

3.5.1 アンテナ・無線通信に関する表現
Expressions for Antennas and Wireless Communications

ここでは，電磁界計算ソフトウェアとマイクロストリップアンテナに関する例文を示します．

There are many computer codes that have been developed to analyze wire-type linear antennas, such as the dipole, and they are too numerous to mention here. One simple program to characterize the radiation characteristics of a dipole, designated as Dipole (both in FORTRAN and MATLAB), is included in the attached CD. Another much more advanced program, designated as the Numerical Electromagnetics Code (NEC), is a user-oriented software developed by Lawrence Livermore National Laboratory. It is a Method of Moments (MoM) code for analyzing the interaction of electromagnetic waves with arbitray structures consisting of conducting wires and surfaces. It is probably the most widely distributed and used electromagnetics code (C. A. Balanis, Antenna Theory: analysis and design, p. 214, ⓒ2005 John Wiley & Sons, Inc. John Wiley & Sons, Inc. による許諾を得て転載)．

ダイポールアンテナのような線状アンテナを解析するために開発され

た計算機コードは非常に多くあります．あまりにも多くて，ここでは述べられません．ダイポールアンテナの放射特性を明らかにするための，Dipole という名の(FORTRAN と MATLAB 両言語の)簡単なプログラムが添付の CD に収録されています．ほかの，これよりはるかに高度な，数値電磁コード(NEC)という名のプログラムはローレンスリバモア国立研究所で開発された利用者指向のソフトウェアです．このプログラムは，電磁波と線状および面状導体から構成される任意構造物との相互作用を解析するためのモーメント法(MoM)に基づく計算機コードです．それは，おそらく最も広く行き渡り，用いられている電磁コードです．

〈補足〉 "computer code" は「計算機コード」，"wire-type linear antenna" は「線状アンテナ」，"user-oriented software" は「利用者指向のソフトウェア」，"method of moments (MoM)" は「モーメント法(MoM)」，"electromagnetic wave" は「電磁波」，"interaction" は「相互作用」です．

Microstrip antennas received considerable attention starting in the 1970s, although the idea of a microstrip antenna can be traced to 1953 and a patent in 1955. Microstrip antennas consist of a very thin ($t \ll \lambda_0$, where λ_0 is the free-space wavelength) metallic strip (patch) placed a small fraction of a wavelength ($h \ll \lambda_0$, usually $0.003\lambda_0 \leq h \leq 0.05\lambda_0$) above a ground plane. The microstrip patch is designed so its pattern maximum is normal to the patch (broadside radiator). This is accomplished by properly choosing the mode (field configuration) of excitation beneath the patch. End-fire radiation can also be accomplished by judicious mode selection. For a rectangular patch, the length L of the element is usually $\lambda_0/3 < L < \lambda_0/2$. The strip (patch) and the ground plane are separated by a dielectric sheet (referred to as the substrate).

There are numerous substrates that can be used for the design of microstrip antennas, and their dielectric constants are usually in the range of $2.2 \leq \varepsilon_r \leq 12$. The ones that are most desirable for good antenna performance are thick substrates whose dielectric constant is in the lower end of the range because they provide better efficiency, larger bandwidth, loosely bound fields for radiation into space, but at the expense of larger element size. Thin substrates with higher dielectric constants are desirable for microwave circuitry because they require tightly bound fields to minimize undesired radiation and coupling, and lead to smaller element sizes; however, because of their greater losses, they are less efficient and have relatively smaller bandwidths. Since microstrip antennas are often integrated with other microwave circuitry, a compromise has to be reached between good antenna performance and circuit design.

Often microstrip antennas are also referred to as patch antennas. The radiating elements and the feed lines are usually photoetched on the dielectric substrate. The radiating patch may be square, rectanglar, thin strip (dipole), circular, elliptical, triangular, or any other configuration. Square, rectangular, dipole (strip), and circular are the most common because of ease of analysis and fabrication, and their attractive radiation characteristics, especially low cross-polarization radiation. Microstrip dipoles are attractive because they inherently possess a large bandwidth and occupy less space, which makes them attractive for arrays. Linear and circular polarizations can be achieved with either single elements or arrays of microstrip antennas. Arrays of microstrip elements, with single or multiple feeds, may also be used to introduce scanning capabilities and achieve greater directivities (C. A. Balanis, Antenna Theory: analysis and design, pp. 812-813, ©2005 John Wiley & Sons, Inc. John Wiley & Sons, Inc. による許諾を得て転載).

　マイクロストリップアンテナの着想は1953年に，そして1955年の特許に端を発しますが，マイクロストリップアンテナは1970年代から相当な注目を集めていました．マイクロストリップアンテナは，接地板から波長のほんのわずかな分(高さ $h \ll \lambda_0$，通常は $0.003\lambda_0 \leq h \leq 0.05\lambda_0$)上に設けた非常に薄い(厚さ $t \ll \lambda_0$，ここで，λ_0 は自由空間波長)金属ストリップ(パッチ)からなります．マイクロストリップパッチは，その放射パターンがパッチ(ブロードサイド放射体)に垂直になるように設計されます．これは，パッチ下の励振モード(電磁界形状)を適切に選ぶことによって実現されます．エンドファイア放射もモードを賢明に選択することにより実現することができます．長方形パッチの長さ L は，通常は $\lambda_0/3 < L < \lambda_0/2$ の範囲です．ストリップ(パッチ)と接地板は誘電体シート(基板と呼ばれる)で隔てられます．

　マイクロストリップアンテナの設計に使用できる基板はたくさんあり

ます．それらの誘電率は，通常は $2.2 \leq \varepsilon_r \leq 12$ の範囲にあります．優れたアンテナ性能のために最も望ましいものは，この範囲の下限の誘電率をもつ厚い基板です．なぜなら，それらは，素子サイズがより大きくなるという犠牲を払いますが，よりよい効率，より広い帯域，空間への放射に適した緩やかな結合をもたらすからです．マイクロ波回路は不要な放射や結合を最小限にするため，強く束縛された電磁界を必要としますので，より高い誘電率をもつ薄い基板はマイクロ波回路にとっては望ましく，小形化につながります．しかし，それらは，より大きな損失をもつため，効率はより低くなり，比較的狭い帯域となります．マイクロストリップアンテナは，多くの場合，ほかのマイクロ波回路と一体化していますので，優れたアンテナ性能と回路設計の間で，妥協点が見出されなければなりません．

　マイクロストリップアンテナはパッチアンテナとも呼ばれることも頻繁にあります．放射素子と給電線は，通常，誘電体基板上で光エッチングされます．放射パッチは，正方形，長方形，細長い一片（ダイポール），円形，だ円形，三角形，あるいはそのほかの形状です．解析および製作が容易なこと，魅力的な放射特性をもつこと，特に交差偏波放射が低いことから，正方形，長方形，ダイポール（ストリップ），そして円形が最も普及しています．マイクロストリップダイポールは魅力的です．なぜなら，それらは本質的に広帯域であり，多くの空間を占有しないからです．これらのことは，マイクロストリップアンテナをアレイとして魅力的にします．直線および円偏波は，マイクロストリップ単一素子あるいはアレイによって実現されます．単一あるいは複数の給電部をもつマイクロストリップ素子のアレイは，走査能力を取り入れるためや，より大きな指向性を達成するために用いられるかもしれません．

〈補足〉 "microstrip antenna" は「マイクロストリップアンテナ」，"receive attention" は「注目を受ける（集める）」，"in the 1970s" は「1970 年代に」，"can be traced to..." は「...に端を発する」，"patent" は「特許」，"free-space wavelength" は「自由空間（における電磁波の）波長」，"a small fraction of..." は「ほんのわずかの...」，"normal to..." は「...に垂直な」，"referred to as..." は「...と呼ばれる（称される）」，"substrate" は「基板」，"dielectric constant" は「誘電率」，"at the expense of..." は「...を犠牲にして」，"circuitry" は「回路」，"be integrated with..." は「...と一体化している」，"feed line" は「給電線」，"elliptical" は「だ円形の」，"fabrication" は「製作」，"linear and circular polarizations" は「直線および円偏波」，"scanning capability" は「走査能

3.5 情報通信に関する表現

力」，"directivity" は「指向性」です．

3.5.2 光ファイバ通信に関する表現
Expressions for Optical Fiber Communications

ここでは，光ファイバ通信に関する例文を示します．

The success of fiber optics has proven beyond reasonable doubt that optics is the best broad-band and long-haul communication medium. The momentum continues as optical communication technologies are penetrating into local-area networking markets. As the application range of optical communications widens from long-distance telecommunications to short-distance data communications, research and development efforts have looked ahead for even shorter distance applications: optical interconnections among intracomputer components at backplane, board, and even chip levels. Although optics maintains its high bandwidth and noise-immunity advantages over electronics, concerns about large space consumption, uses of additional media conversion, and lack of packaging choices inhibit many practical applications. On the other hand, electronics is gradually approaching its limits as far as the fundamental physics and cost-effectiveness are concerned.

Technological advances in fabricating a low-power, high-efficiency vertical-cavity surface-emitting laser (VCSEL) are leading the way to push the door of mainstream applications of optoelectronics (OE) to computing. Integration efforts are under way to combine VCSELs and photodetectors and their basically functional circuits together to form the so-called smart-pixel arrays. The availability of such smart-pixel arrays will be crucial to the success of optoelectronics active component technology as far as optical interconnects are concerned. To successfully use OE interconnections, technologies to fabricate, integrate, and package massive optical transmission channels to handle large-bandwidth, low-crosstalk parallel optical data are also crucially important (Li et al., p. 794, ©2000 IEEE. IEEEによる許諾を得て転載)．

光ファイバの成功により，光ファイバが最もよい広帯域長距離通信媒体であることが，合理的疑いの余地なく証明されました．光通信技術が

ローカルエリアネットワーク市場に進出しているように，その勢いは続いています．光通信の応用範囲は長距離電話通信から短距離データ通信まで広がっており，さらなる短距離通信への応用となる背面配線板，基板，集積回路レベルでのコンピュータ内部要素間の光配線を見据えて研究開発の努力がなされてきました．光通信機器は，電子通信機器に比べて，広帯域であり，耐雑音性が高いという利点をもっていますが，広い空間を消費すること，追加の媒体変換を使用すること，パッケージングの選択肢が不足していることについての懸念が，多くの実用的応用を妨げています．一方で，電子通信機器は，その基本的な物理的特性およびコストパフォーマンスに関するかぎり，その限界に近づいています．

　低電力，高効率の垂直キャビティ面発光レーザ(VCSEL)の製造技術の進歩は，光電子工学(OE)からコンピューティングに光ファイバ応用の本流が移行するのを先導しています．いわゆるスマートピクセルアレイを形成するために，VCSELと光センサとそれらの基本機能回路を一体化するための統合に向けた取組みが進行中です．このようなスマートピクセルアレイを利用できることが，光配線に関するかぎり，光電子能動部品技術の成功に不可欠です．光電子接続をうまく使用すること，広帯域，低クロストークの並列光データを取り扱うための多数の光伝送路を製造し，統合し，そして包装する技術もまた非常に重要です．

〈補足〉 "fiber optics"は「光ファイバ」，"broadband"は「ブロードバンド(広帯域)」，"long-haul"は「長距離の」，"optical communication"は「光通信」，"local area network"は「ローカルエリアネットワーク」，"optical interconnection"は「光配線(接続)」，"back plane"は「背面配線板」，"noise immunity"は「耐雑音性」，"cost-effectiveness"は「コストパフォーマンス(費用対効果)」，"vertical-cavity surface-emitting laser (VCSEL)"は「垂直キャビティ面発光レーザ(VCSEL)」，"optoelectronics"は「光電子工学」，"photodetector"は「光センサ」，"so-called"は「いわゆる」，"smart-pixel array"は「スマートピクセルアレイ」，"crosstalk"は「クロストーク(混線，漏話)」です．
　なお，「光ファイバ」を"optical fiber"ともいいます．

After conquering the core and metropolitan networks, optical fiber is now penetrating into the access domain. Its low loss and huge bandwidth enable the delivery of any present and foreseeable set of broadband services, and also make it a nice match to the wireless link to the end user. Cost effectiveness is a key issue, and will be decisive for the network topology choices. Point-to-point may be the most cost-effective for short-reach access, whereas point-to-multipoint may be the most interesting at medium-to long-reach access, or when line terminations in the local exchange become a key issue. A number of optical techniques being deployed for shared-fiber multiple access are discussed, on the basis of time slot multiplexing, frequency slot multiplexing, code division multiplexing, and wavelength multiplexing, including their application in fiber to the home/fiber to the premises (FTTH/FTTP) networks for fast data transfer (asynchronous transfer mode (ATM) or Ethernet based) and for broadband service distribution (such as CATV) (Koonen, p. 911, ©2006 IEEE. IEEE による許諾を得て転載).

光ファイバは，基幹および都市回線網を制圧し，今やアクセスドメインに浸透しつつあります．その低損失と非常に広い周波数帯域は，現在および当面の一連のブロードバンドサービスの配信を可能にします．また，エンドユーザへの無線リンクにもよく適合します．コストパフォーマンスが重要な課題で，それがどのネットワークの構成形態を選択するかを決めることになります．ポイントツーポイントが，短距離アクセスではおそらく最もコストパフォーマンスが高いでしょう．一方，中から長距離アクセスにおいて，あるいはローカル交換網での回線端末が重要な課題となる場合には，ポイントツーマルチポイントが最も関心を引くでしょう．時間スロット多重化，周波数スロット多重化，符号分割多重化，波長分割多重化，および高速データ転送（非同期転送モード（ATM）またはイーサネットに基づく）のための，そしてブロードバンドサービス配信（CATVのような）のための，それらのFTTH/FTTP回線網における応用に基づき，光ファイバ共有多重アクセスに用いられている多くの光学技術が論じられています．

〈補足〉 "core network"は「基幹回線網」，"end user"は「エンドユーザ（末端使用者）」，"network topology"は「ネットワークトポロジー（ネットワークの構成形態）」，"point-to-point"は「ポイントツーポイント（1地点対1地点）」，"point-to-multipoint"は「ポイントツーマルチポイント（1地点対多地点）」，"multiplexing"は「多重化」，"code division multiplexing"は「符号分割多重化」，"fiber to the home (FTTH)"は「各家庭までの光ファイバ化（FTTH）」，"fiber to the premises (FTTP)"は「各敷地までの光ファイバ化（FTTP）」，"asynchronous transfer mode (ATM)"は「非同期転送モード（ATM）」，

"Ethernet"は「イーサネット」，"cable television (CATV)"は「ケーブルテレビ(CATV)」です．

> In a time division multiple access (TDMA) system, the upstream packets from the optical network units (ONUs) are time-interleaved at the power splitting point, which requires careful synchronization of the packet transmission instants at the ONUs. This synchronization is achieved by means of grants sent from the local exchange, which instruct the ONU when to send a packet. The correct timing of these submissions is achieved by ranging protocols, which sense the distance from each ONU to the local exchange. In the optical line terminal (OLT) at the local exchange, a burst mode receiver is needed which can synchronize quickly to packets coming from different ONUs, and which also can handle the different amplitude levels of the packets due to differences in the path loss experienced. As the ONUs are sharing jointly the capacity of the OLT, the average capacity per ONU decreases when the number of ONUs grows (Koonen, p. 914, ©2006 IEEE. IEEEによる許諾を得て転載)．
>
> 時分割多重アクセス(TDMA)システムにおいては，光回線終端装置(ONU)からのアップストリームパケットは，パワー分岐点で時間インタリーブされます．このためには，ONUでのパケット伝送の瞬間を注意深く同期する必要があります．この同期は，ローカル交換局から送信されるグラント信号により実現されます．これらを発信するための正しいタイミングは，各ONUからローカル交換局までの距離を識別するレンジングプロトコルによりとられています．ローカル交換局の光通信回線端末(OLT)では，異なるONUからのパケットにすぐに同期することができ，また経路損失の相違によるパケット振幅の相違に対応できるバーストモード受信器が必要です．ONUはOLTの容量を共有しているため，ONUの数が増えるとONU1機あたりの平均容量は減少します．

〈補足〉 "time division multiple access (TDMA)"は「時分割多重アクセス」，"optical network unit (ONU)"は「光回線終端装置」，"upstream packet"は「アップストリームパケット(上流の通信機器に送られるパケット)」，"time-interleave"は「時間インタリーブする」，"synchronization"は「同期すること」です．

3.5.3 インターネットに関する表現
Expressions for the Internet

ここでは，インターネットに関する例文を示します．

In 1999, phone and cable companies in North America, Europe, and other industrialized regions put into place the technologies needed to dramatically boost Internet access speeds—that is, digital subscriber loop (DSL), cable modems, and satellite links—and started hooking people up en masse. These new connections are transforming the Internet experience. For one thing, they do away with the so-called World Wide Wait, now that Web pages can stream in over 100 times faster than with 56-kb/s modems. For another, they put everyone on-line whenever they fire up their systems, letting e-mail come in and go out as easily as phone calls. It is a good thing that the speed is increasing, because new developments at the World Wide Web Consortium, Cambridge, Mass., are going to require more data and make the Web more appealing. The consortium is changing the look and feel of the Web, enriching it both sensually and in terms of data usage.

The change embodies two new recommendations, as the consortium terms standards: the extended hypertext markup language and the synchronized multimedia integration language (XHTML and SMIL, respectively). Together, they alter how Web documents are created, in the process making it much easier to use more content, or data, in each Web page.

XHTML goes on from where earlier versions of the HTML language, more especially HTML 4.0, left off. It completely overhauls HTML to make it compatible with the extensible markup language (XML), a simplified version of the standard graphics markup language (SGML). The result of all this work is that XHTML users will find it easy to specify the precise look of a Web page (which is what SGML did for electronic documents, but in a very complex way). Pictures will appear where they should, not on top of text, and text will have the kind of font, leading, and justification the designer wants. So designers will be free to use more graphics and assorted fonts.

SMIL was released as version 1.0 in 1998, but was hard to use. So a new version is being readied for release in midyear. Once deployed, the new version will let designers easily enrich pages with multimedia and, like XHTML, will give them more control of a Web page's appearance. Suppose the goal is a Web page in which an animation sequentially brings up pictures of popular 1950s children's heros: The Lone Ranger, Superman, Batman, and so on. As each appears, the designer would like the hero's theme music to play (The William Tell Overture, for instance, would accompany the picture of the Lone Ranger). Today, the designer would have to attempt the well-nigh impossible—add just enough dead air to the start of the audio file to give the animation time to download. Of course, since download times vary widely, the designer would have to be omniscient to know how much dead air to add (Comerford, pp. 40-41, ©2000 IEEE. IEEEによる許諾を得て転載).

　1999年に，北アメリカ，ヨーロッパおよびほかの先進工業地域における電話およびケーブル会社は，インターネットのアクセス速度を飛躍的に上げるために必要な技術〔つまり，ディジタル加入者回線(DSL)，ケーブルモデム，および衛星中継〕を整えました．そして，一斉に人々をとりこにし始めました．これらの新しい接続はインターネット経験を変えています．一例をあげると，それらは，いわゆるワールドワイドウェイト(世界的な待ち時間)をなくし，今やウェブページは，56 kb/sのモデムを用いていた際の100倍以上の速度で配信することができます．別の例では，システムが起動するといつでも皆をインターネットに接続し，電話のように簡単に電子メールをやり取りさせます．マサチューセッツ州ケンブリッジのワールドワイドウェブコンソーシアムでの新展開はより多くのデータを必要とするでしょうし，ウェブをより魅力的なものにするでしょうから，アクセス速度が上がっていくことはよいことです．このコンソーシアムは，ウェブのイメージを変え，ウェブを感覚的にもデータ利用の観点からも強化しつつあります．

　その変化は，コンソーシアムが標準と呼ぶ二つの提言〔拡張したハイパーテキストマークアップ言語と同期化マルチメディア統合言語(それぞれ，XHTMLとSMIL)〕を具体化しています．それらはともに，各ウェブページにおいてより多くの内容またはデータを使うのをより簡単にする過程において，ウェブ文書の作り方を変えます．

　XHTMLは，HTML言語の旧バージョン，特にHTML 4.0の続きです．それは，標準グラフィックスマークアップ言語(SGML)の簡略版で

ある規約拡張マークアップ言語(XML)と互換性をもたせるために，HTML を徹底的に見直しています．この成果は，XHTML ユーザがウェブページの体裁を緻密に設定するのを容易にするでしょう（これは，SGML が電子文書に対してしたことなのですが，非常に複雑なやり方でした）．写真は，文字列の上ではなく，しかるべきところに現れ，文字列は，デザイナの望むフォント，行送りおよび行端ぞろえをもつでしょう．したがって，デザイナは，自由により多くの画像やさまざまなフォントを使えるでしょう．

SMIL は 1998 年にバージョン 1.0 としてリリースされましたが，使いづらいものでした．このため，新バージョンが年の中ごろにリリースされるように準備されています．新バージョンが使えるようになると，デザイナたちはマルチメディアを用いてページを容易に充実させるでしょうし，XHTML を用いた場合のように，ウェブページの外観をもっと操作するでしょう．もしも，その目的が，人気のある 1950 年代の子供たちのヒーロー（ローンレンジャー，スーパーマン，バットマンなど）の写真を連続して繰り出すウェブページであれば，デザイナは，それぞれの写真が現れるときに，そのヒーローのテーマ曲を流したいと思うでしょう（たとえば，ウィリアム・テルの序曲がローンレンジャーの写真に伴うでしょう）．今日では，デザイナはほとんど不可能なこと（そのアニメーションにダウンロードする時間を与えるため，音声ファイルの開始に過不足のないむだ時間を加えること）を試みなければなりません．もちろん，ダウンロード時間は大きく変るため，どの程度のむだ時間を加えるべきかをよく知っていなければなりません．

〈補足〉 "digital subscriber loop or line (DSL)" は「ディジタル加入者回線」，"satellite link" は「衛星中継」，"hook" は「とりこにする（接続する）」，"en masse" は「（フランス語）一斉に」，"the World Wide Web Consortium" は「ワールドワイドウェブコンソーシアム」，"extended hypertext markup language (XHTML)" は「拡張したハイパーテキストマークアップ言語 (XHTML)」，"synchronized multimedia integration language (SMIL)" は「同期化マルチメディア統合言語(SMIL)」，"extensible markup language (XML)" は「規約拡張マークアップ言語(XML)」，"leading" は「行送り」，"justification" は「行端ぞろえ」，"well-nigh" は「ほとんど」です．

The technology underlying the Web must also change if it is to keep up with demand. For one thing, the 32-bit Internet Protocol (IP) addresses specified by IP version 4 will soon be exhausted, already they are in short supply. The only cure will be to roll out the next version of the protocol, IPv6, with its 128-bit address space. While the new standard has been ready for some time, there has been little need for it. So network administrators have deferred the switch to software that supports it, which in turn prevents use of the longer numerical addresses.

Growing the address space is not the only fix needed. Today, the Internet works by having each router, or network node, try its best to deliver each packet of data, and treat all data packets as equal. So packets may be delivered randomly in no particular sequence or, even worse, not at all. While out-of-sequence arrival may not matter for electronic files that can be reassembled in a computer, it is disconcerting when it happens during an IP phone call.

For that reason, technologists working on the next generation of the Internet are experimenting with a new standard called differentiated services, or DiffServ. Created within the Internet Engineering Task Force, it will allow different kinds of traffic to be handled with different priorities, as well as allow businesses to offer users different types of service at different prices (Comerford 2000, p. 41, ©2000 IEEE. IEEE によ る許諾を得て転載).

　需要についていくためには，ウェブの基本となる技術もまた変化しなければなりません．一例をあげると，IPバージョン4で指定される32ビットのインターネットプロトコル(IP)アドレスはすぐに使い尽くされてしまうでしょう．すでに不足している状況です．唯一の解決策は，次期バージョンである128ビットアドレス空間をもつIPv6を公開することでしょう．その新しい標準はしばらく前に準備ができていましたが，必要性がほとんどありませんでした．したがって，ネットワーク管理者は，それをサポートするソフトウェアへの切換えを先延ばしにしてきました．その結果，長い数値アドレスを使用せずにすんでいます．

　アドレス空間を増やすことが必要な唯一の解決策ではありません．今日，インターネットは，ルータあるいはネットワークノードをもつことによって機能しています．そして，データの各パケットを届けるのに最善を尽くし，すべてのパケットを平等に扱っています．したがって，パケットは，特別な順序なく，あるいは全く順序なく，手当たり次第に届

けられるかもしれません．電子ファイル群はコンピュータで再構築されるので，順序なく届くことは問題ではありませんが，IP 電話中にそれが起こると困ります．

　この理由で，次世代のインターネットについて研究している科学技術者たちは，分化サービスあるいはディフサーブと呼ばれる新標準を用いて実験を行っています．これはインターネット技術委員会で作り出されたもので，これにより，異なる種類のトラフィックを異なる優先度で扱えます．また，会社が異なる価格で異なるタイプのサービスをユーザに提供することもできます．

〈補足〉　"Internet Protocol (IP)" は「インターネットプロトコル (IP)」，"network administrator" は「ネットワーク管理者」，"Task Force" は「委員会（作業部会）」です．

3.6　機械学習・人工知能（AI）に関する表現
Expressions for Machine Learning and Artificial Intelligence

　ここでは，機械学習，深層学習，機械翻訳，生成 AI 分野で用いられる諸表現を示します．

3.6.1　機械学習と深層学習に関する表現
Expressions for Machine Learning and Deep Learning

ここでは，機械学習と深層学習に関する例文を示します．

　People send and receive data across network infrastructure, such as a router, that can be intercepted and manipulated by outsiders. The increased use of the Internet has increased the amount and complexity of data, resulting in the emergence of big data. The constant rise of the Internet and extensive data necessitated the creation of a reliable intrusion detection system. Network security is a subset of cybersecurity that safeguards systems connected to a network against malicious activity. The goal is to provide networked computers to ensure data security, integrity, and accessibility. Current cybersecurity research focuses on creating an effective intrusion detection system that can identify both known and new attacks and threats with high accuracy and a low false alarm rate.

　The terms Artificial Intelligence (AI), Machine Learning (ML), and Deep Learning (DL) are frequently used interchangeably to describe the same principles in software development. These names all indicate the same thing: a machine programmed to learn and find the best solution to a problem. DL is a subfield of machine learning, whereas machine learning is a subfield of AI. As a result, ML and DL are employed to create an efficient and effective intrusion detection system. (Halbouni et al., IEEE Access, vol. 10, pp. 19572-19573, 2022 IEEE).

　部外者によって傍受されたり操作されたりする可能性があるルーターなどのネットワークインフラを介して，人々はデータを送受信します．インターネットの利用の増加により，データの量と複雑さが増し，その結果，ビッグデータが出現しました．インターネットと膨大なデータの絶え間ない増加により，信頼性の高い侵入検知システムの開発が必要になりました．ネットワークセキュリティは，ネットワークに接続されているシステムを悪意のある行為から保護するサイバーセキュリティのサブセットです．その目的は，データのセキュリティ，整合性およびアクセス性を確保するために，ネットワークに接続されたコンピュータに提供することです．現在のサイバーセキュリティの研究は，既知および新規の両方の攻撃と脅威を高精度，低誤警報率で識別できる効果的な侵入検知システムの開発に重点が置かれています．

　人工知能(AI)，機械学習(ML)および深層学習(DL)という用語は，ソフトウェア開発において同じ原理を説明するためにしばしば同義で使用されます．これらの名称はすべて同じこと，つまり，ある問題に対する最適解を学習して見つけるようにプログラムされた機械を示しています．DLは機械学習のサブフィールドですが，機械学習はAIのサブフィールドです．その結果，MLとDLは効率的かつ効果的な侵入検知システムを開発するために用いられます．

〈補足〉 "intrusion detection system" は「侵入検知システム」，"subset" は「サブセット(小集団，部分集合)」，"malicious activity" は「悪意のある行為」，"cybersecurity" は「サイバーセキュリティ(インターネットなどにおいて不正侵入，データの流出，改ざんなどからシステムを保護するための対策)」，

3.6 機械学習・人工知能(AI)に関する表現

"false alarm rate"は「誤警報率」，"artificial intelligence (AI)"は「人工知能(AI)」，"machine learning (ML)"は「機械学習(ML)」，"deep learning"は「深層学習」です．

> Automatic analyses of attacks and security events, such as spam mail, user identification, social media analytics, and attack detection may be performed efficiently using machine learning. There are the following main techniques of machine learning: supervised, unsupervised, semi-supervised, and reinforcement learning. Supervised learning is based on labeled data, unsupervised learning is based on unlabelled data, and semi-supervised learning is based on both.
>
> Traditional machine learning techniques are limited to processing natural raw data that rely on adequate feature extraction, and in order to classify or find patterns by a classifier, the raw data must be transformed into the appropriate format, which is where deep learning comes in. Deep learning is a machine learning approach that can learn from unstructured or unlabeled data and representation based on human brain knowledge.
>
> Deep learning is motivated by neural networks (NN), which can mimic the human brain and perform analytical learning by analyzing data like text, images, and audio. In contrast to deep learning models, which feature multiple connected layers, shallow learning models are built up of a few hidden layers. By stacking layers on top of layers, DL will be able to express increasing complexity functions more effectively. DL is used to learn representations with many abstraction levels. Deep neural networks are capable of finding and learning representations from raw data and performing feature learning and classification. Machine learning methodologies are also utilized in deep learning (Halbouni et al., IEEE Access, vol. 10, p. 19573, 2022).
>
> スパムメール，ユーザー識別，ソーシャルメディア分析，攻撃検出などの攻撃やセキュリティ事象の自動分析は，機械学習を用いて効率的に実行できる可能性があります．機械学習には，教師あり学習，教師なし学習，半教師あり学習，強化学習という主要な手法があります．教師あり学習はラベル付きデータに基づいており，教師なし学習はラベルなしデータに基づいており，半教師あり学習は両方に基づいています．

従来の機械学習技術は，適切な特徴抽出に依存する未加工データの処理に限定されており，分類子によってパターンを分類または検索するには，未加工データを適切な形式に変換する必要があり，そこに深層学習が現れます．深層学習は，人間の脳の知識に基づいた構造化されていないあるいはラベル付けされていないデータや表現から学習できる機械学習法です．

深層学習は，人間の脳を模倣し，テキスト，画像，音声などのデータを分析して学習することができるニューラルネットワーク(NN)によって動かされています．複数の接続された層を特徴とする深層学習モデルとは対照的に，浅層学習モデルはいくつかの隠れ層で構築されています．層の上に層を積み重ねることにより，深層学習は複雑さが増す関数をより効果的に表現できるようになります．深層学習は，多くの抽象化レベルをもつ表現を学習するために使用されます．深層ニューラルネットワークは，未加工データから表現を見つけて学習し，特徴の学習と分類を実行することができます．機械学習手法は深層学習でも利用されます．

〈補足〉 "supervised learning"は「教師あり学習」，"unsupervised learning"は「教師なし学習」，"semi-supervised learning"は「半教師あり学習」，"reinforcement learning"は「強化学習」，"neural network (NN)"は「ニューラルネットワーク(NN)」，"mimic"は「模倣する」，"hidden layer"は「隠れ層」です．

3.6.2　機械翻訳に関する表現
Expressions for Machine Translation

ここでは，機械翻訳に関する例文を示します．

In recent years, machine translation (MT) has become essential in many applications and has achieved advances for almost all languages. Consequently, the recent progress of MT has boosted the translation quality significantly, and even approached human translation quality for high-resource language pairs like English-Czech or Chinese-English. Likewise, other research subfields in Natural Language Processing (NLP) systems also achieved a great improvement aiming to satisfy the MT growing needs in several domains and applications such as in multilingual chatbots. Therefore, the demand for quick and accurate machine translations is growing accordingly. However, finding appropriate and optimal translation is not an easy task in any language setting. Several machine translation systems already exist, such as Sakhr, Al-Mutarjim^{TM} Al-Arabey 3.0 and SYSTRAN, but the quality of the translation with regards to the Arabic language needs to be further improved due to many issues reported in various research works such as linguistic errors. In addition, there are web-based MT systems such as Babylon, Bing Translator, Google Translate, and Shaheen that can translate a source text from Arabic to English and vice-versa（Zakraoui et al., IEEE Access, vol. 9, p. 161445）.

　近年，機械翻訳(MT)は多くのアプリケーションで不可欠なものとなり，ほぼすべての言語で進歩を遂げています．その結果，最近のMTの進歩により翻訳品質が著しく向上し，英語－チェコ語や中国語－英語のような高リソース言語ペアの翻訳品質は人間による翻訳品質にさえも近づいています．同様に，自然言語処理(NLP)システムの他の研究サブ分野も，多言語チャットボットのようないくつかの領域や応用分野で増大する機械翻訳のニーズを満たすことを目的として，大幅に向上しました．したがって，迅速かつ正確な機械翻訳に対する需要もそれに応じて高まっています．しかし，適切かつ最適な翻訳を見つけることは，どの言語設定においても容易な作業ではありません．Sakhr，Al-Mutarjim^{TM} Al-Arabey 3.0およびSYSTRANのような機械翻訳システムはすでに存在していますが，さまざまな研究論文で言語上の誤りなど多くの問題が報告されているため，アラビア語に関する翻訳の品質はさらなる向上が必要です．それ加えて，ソーステキストをアラビア語から英語に，そしてその逆に翻訳できるBabylon，Bing Translator，Google TranslateおよびShaheenなどのウェブベースの機械翻訳システムもあります．

〈補足〉 "machine translation"は「機械翻訳」，"high-resource language"は「高リソース言語(学習用データが豊富な言語)」，"natural language processing"は「自然言語処理(人間の言語をコンピュータで処理し内容を抽出すること)」，"multilingual"は「多くの言語を使える」，"chatbot"は「チャットボット(自動会話プログラム)」です．

Machine Translation (MT) is a computer application that translates texts or speech from one natural language (i.e., the source) to another (i.e., the target). MT generates a sentence, by translating a source sentence, that gives its meaning in the target language. The translation process which deals with semantic, syntactic, morphological, and additional varieties of grammatical complexities of both languages is hard and complex. In fact, the problem is harder in languages where the source and the target languages have a wide array of linguistic dissimilarities. For example, the Arabic language differs from the target language, such as English, at the phonological, orthographical, morphological, syntactical, and lexical levels.

First, phonologically, Arabic contains 28 consonants, 3 short vowels, 3 long vowels, 2 diphthongs. Arabic spelling is mostly phonemic corresponding to its letter sound. Second, orthographically, the Arabic script is an alphabet with allographic variants, optional zero-width diacritics, and common ligatures. The shape of a single Arabic letter may change slightly depending on its position within the Arabic word (beginning, middle, or at the end). Third, morphologically, Arabic morphology has two main functions, notably inflection and derivation. Inflectional function, which is mostly concatenative, modifies features of words such as tense, number, person, etc. Derivational function, which is mostly templatic, creates new words by inserting one or multiple affixes. Fourth, syntactically, Arabic syntax is heavily related to its morphological level. Thus, several syntactic aspects are not expressed uniquely via word order but also through morphology. The Arabic admits two types of sentences: verbal and nominal. Arabic is case marking: nominative, accusative, genitive, and almost-free word order.(Zakraoui et al., IEEE Access, vol. 9, p. 161447, 2021).

機械翻訳(MT)は，テキストまたは音声を一つの自然言語(ソース)から別の自然言語(ターゲット)に翻訳するコンピュータアプリケーションです．MTは，ソース文を翻訳することによって，ターゲット言語でその意味を与える文を生成します．両言語の意味論的，構文的，形態論的，およびその他のさまざまな文法上の複雑さを扱う翻訳プロセスは，難解で複雑です．実際，ソース言語とターゲット言語に幅広い言語的相違が

ある言語では，この問題はより難しくなります．例えば，音韻論的，正書法的，形態論的，構文的，語彙的レベルにおいて，アラビア語は英語などのターゲット言語とは異なります．

　まず，音韻的には，アラビア語には28個の子音，3個の短母音，3個の長母音，2個の二重母音が含まれています．アラビア語の綴りは，ほとんどが文字の音に対応する音素です．第二に，正書法的には，アラビア文字は異文字異綴語，任意のゼロ幅付加記号，および一般的な合字を備えた文字体系です．単一のアラビア文字の形状は，アラビア語内の位置（先頭，中間または末尾）に依存して，わずかに変化することがあります．第三に，形態論的に，アラビア語の形態論には二つの主な機能，特に屈折と派生があります．屈折の機能は，ほとんどの場合連結であり，時制，数字，人称などの単語の特徴を変えます．派生の機能は，ほとんどの場合定型であり，一つまたは複数の接辞を挿入することによって新しい単語を作成します．第四に，構文的には，アラビア語の構文はその形態学的レベルに大きく関係しています．したがって，いくつかの構文的側面は，語順だけでなく形態論によっても一意的に表現されます．アラビア語では，口頭と名目上の二種類の文が認められます．アラビア語は格標示です（主格，対格，所有格とほぼ自由な語順です）．

〈補足〉　"semantic"は「意味論的な」，"syntactic"および"syntactical"は「構文的な」，"morphological"は「形態論的な」，"phonological"は「音韻論的な」，"orthographical"は「正書法的な」，"lexical"は「語彙的な」，"consonant"は「子音」，"vowel"は「母音」，"diphthong"は「二重母音」，"phonemic"は「音素」，allographic"は「異文字の」，"variant"は「異綴語」，"diacritic"は「付加記号」，"ligature"は「合字（二つも文字が接続されたもの）」，"inflection"は「屈折」，"derivation"は「派生」，"concatenative"は「連結的な」，"affix"は「接辞」，"case marking"は「格標示」，"nominative"は「主格」，"accusative"は「対格」，"genitive"は「所有格」です．

3.6.3　生成AIに関する表現
Expressions for Generative Artificial Intelligence

ここでは，生成AIに関する例文を示します．

The evolution of Artificial Intelligence (AI) and Machine Learning (ML) has led the digital transformation in the last decade. AI and ML have achieved significant breakthroughs starting from supervised learning and rapidly advancing with the development of unsupervised, semi-supervised, reinforcement, and deep learning. The latest frontier of AI technology has arrived as Generative AI (GenAI). GenAI models are developed using deep neural networks to learn the pattern and structure of big training corpus to generate similar new content. GenAI technology can generate different forms of content like text, images, sound, animation, source code, and other forms of data. The launch of ChatGPT (Generative Pre-trained Transformer), a powerful new generative AI tool by OpenAI in November 2022, has disrupted the entire community of AI/ML technology. ChatGPT has demonstrated the power of generative AI to reach the general public, revolutionizing how people perceive AI/ML. At this time, the tech industry is in a race to develop the most sophisticated Large Language Models (LLMs) that can create a human-like conversation, the result of which is Microsoft's GPT model, Google's Bard, and Meta's LLaMa. GenAI has become a common tool on the Internet within the past year. With ChatGPT reaching 100 million users within two months of release, suggesting that people who have access to the Internet have either used GenAI or know someone who has(Gupta et al., IEEE Access, vol. 11, pp. 80218-80219, 2023).

　人工知能(AI)と機械学習(ML)の進化は，過去10年間のデジタルトランスフォーメーションを先導してきました．AIとMLは，教師あり学習から始まり，教師なし学習，半教師あり学習，強化学習および深層学習の開発で急速に進み，飛躍的な進歩を遂げてきました．AI技術の最先端領域は，生成AIとして登場しました．生成AIモデルは，同種の新しいコンテンツを生成するために，大規模なトレーニングコーパスのパターンと構造を学習するための深層ニューラルネットワークを用いて開発されます．生成AI技術は，文章，画像，音声，動画，ソースコードおよびその他の形式のデータなど，さまざまな形式のコンテンツを生成することができます．2022年11月にOpenAI社により開発された強力な新しい生成AIツールであるChatGPT(生成可能な事前学習済み変換器)の登場は，AI/ML技術コミュニティ全体に変革をもたらしました．ChatGPTは，一般人に届く生成AIの力を実証し，人々のAI/MLに対する認識に革命を起こしました．現在，技術産業界は，人

間のような会話を作成できる最も洗練された大規模言語モデル（LLM）の開発競争を行っています．その結果がMicrosoft 社のGPT モデル，Google 社のBard，Meta 社のLLaMa です．生成AI は，過去1年でインターネット上での一般的なツールになりました．ChatGPT はリリースから2か月以内にユーザーが1億人に達しており，このことはインターネットにアクセスした人は生成AI を使用したことがある，あるいは生成AI を使用した人を知っていることを示しています．

〈補足〉 "digital transformation" は「デジタルトランスフォーメーション」，"generative AI" は「生成AI」，"corpus" は「コーパス（自然言語の文章を大量に集めて整理したデータベース）」，"source code" は「ソースコード（プログラミング言語で書かれたプログラム）」，"Meta" は「Meta 社（旧 Facebook 社）」です．

The history of generative models dates back to the 1950s when Hidden Markov Models (HMMs) and Gaussian Mixture Models (GMMs) were developed. The significant leap in the performance of these generative models was achieved only after the advent of deep learning. One of the earliest sequence generation methods was N-gram language modeling, where the best sequence is generated based on the learned word distribution. The introduction of Generative Adversarial Network (GAN) significantly enhanced the generative power from these models. The latest technology that has been the backbone of much generative technology is the transformer architecture, which has been applied to LLMs like BERT and GPT. GenAI has evolved in numerous domains like image, speech, text, etc. However, we will only be discussing text-based AI chatbots and ChatGPT in particular relevant to this work. Since ChatGPT is powered by GPT-3 language model, we will briefly discuss the evolution of the OpenAI's GPT models over time (Gupta et al., IEEE Access, vol. 11, p. 80219, 2023).

生成モデルの歴史は，隠れマルコフモデル（HMM）と混合ガウスモデル（GMM）が開発された1950 年代に遡ります．これらの生成モデルの性能の大飛躍は，深層学習が出現して初めて達成されました．最も初期の系列生成方法の一つはN-gram 言語モデリングであり，そこでは学習された単語の分布に基づいて最適な系列が生成されます．敵対的生成

ネットワーク(GAN)の導入により，これらのモデルからの生成力は大幅に強化されました．多くの生成技術の骨格となっている最新技術は，BERTやGPTなどの大規模言語モデルに適用されているTransformerアーキテクチャです．生成AIは，画像，音声，テキストなどのような多くの領域で進化してきました．しかし，この言語モデル開発に特に関還するテキストベースのAIチャットボットとChatGPTについてのみ説明します．ChatGPTはGPT-3言語モデルを装備しているため，OpenAI社のGPTモデルの時間経過に伴う進化について簡単に説明します．

〈補足〉 "hidden Markov model (HMM)"は「隠れマルコフモデル(確率モデルの一つ)」，"gaussian mixture model (GMM)"は「混合ガウスモデル(クラスタリング法の一つ)」，"N-gram language model"は「N-gram言語モデル(マルコフ過程に基づき単語列をモデル化する言語モデル)」，"generative adversarial network (GAN)"は「敵対的生成ネットワーク(教師なし学習で使用される人工知能アルゴリズムの一つ)」，"BERT"は「言語モデルBERT (bidirectional encoder representations from transformersの頭字語)，"transformer"は「深層学習モデルTransformer(自然言語処理分野で使われるモデル)」，"chatbot"は「チャットボット(自動会話プログラム)」，"powered by ..."は「...を装備した」です．

付 録

Appendices

付録1　論文投稿から掲載または発表にいたるまでの流れ
付録2　インパクトファクタについて
付録3　h指数(h-index)について
付録4　電子メール文例
付録5　アメリカ留学関連手続き
付録6　電気学会誌に掲載された英語論文執筆,発表法に関する
　　　　文献リスト

付　録

付録 1　論文投稿から掲載または発表にいたるまでの流れ
General Flow from the Submission of a Scientific Paper to its Publication or Presentation

付 1.1　学術雑誌への論文投稿から掲載にいたるまでの流れ
General Flow from the Submission of a Paper to a Scientific Journal to its Publication

　本付録では，学術雑誌に論文を投稿してから，それが掲載されるまでに必要な手続きの一般的な流れを示します．ある特定のトピックに関する特集が組まれた号に投稿する場合以外は，論文投稿は随時受け付けられるので，決まった締切日はありません．**付表 1.1** に，一般的な手続きの流れを示します．

　論文の投稿先には，論文内容に適合する分野で，かつ高い評価を得ている雑誌を選びましょう．高い評価を得ているか否かの指標は，その研究分野の多くの研究者や技術者が所属している学会の雑誌であることや，高いインパクトファクタ（**付録 2** で詳しく述べています）を有している雑誌であることなどです．このような雑誌では，同じ専門分野における見識の高い方々に査読してもらえる可能性が高く，それにより，論文をよりよいものにするために役立つコメントをもらえる可能性も高くなります．また，著者の気づかなかった誤りを指摘してもらえる可能性も高いため，誤った内容を含んだまま投稿論文が掲載されることも避けられます．もしも，第 1 希望の最初に投稿した雑誌で，掲載不可の判定がでた場合には，その際にもらったコメントを参考に論文を修正し，完成度を高め，第 2 希望の雑誌に投稿することになります．ある雑誌に投稿中に，それと同じ成果を別の雑誌に投稿することはできません．

付表 1.1　学術雑誌への論文投稿から掲載にいたるまでの流れ

順序	およその時期（投稿時を起点）	必要手続き，学会事務局などからの連絡	説　　明
1	0 カ月	論文投稿 (Submission of an	まずは，投稿希望の学会または雑誌のウェブページから，投稿規定(Guide for authors または Instructions to authors)を

		original manuscript)	入手しましょう．投稿規定には，その雑誌が対象とする分野，投稿論文フォーマット(要旨の語数や論文のトータルのページ数などに制限がある場合があります)などの情報が記載されています．次に，指定された論文フォーマットあるいはテンプレートに従って作成した論文を，学会または雑誌の論文投稿用ウェブページから電子投稿します．雑誌によっては，論文(Full papers．通常6ページ程度以上)，レターまたは短論文(Letters or short papers．通常4ページ以内)などの論文タイプを選べるものもありますし，そのどちらかのみを受け付けるものもあります．
2	1〜3カ月	査読期間 (Review period)	投稿された論文は，投稿後1〜2週間の間に，編集長(Editor-in-chief)または編集委員(Associate editors)により，2〜4名の査読者(Reviewers)に割り当てられます．すべての査読者から，査読結果が編集委員および編集長に届くのには，数カ月の時間がかかることもあります．投稿後3ヶ月以上，なんの連絡もない場合には，編集長または編集委員に状況確認の電子メールを送りましょう(付4.1に電子メールの文例を示しています)．
3	1〜3カ月	査読結果通知 (Notification of decision on the manuscript)	編集長から論文の査読結果と査読者のコメントを知らせる電子メールが届きます．編集長は，査読結果に基づき，"Accept"(掲載受付)，"Revise and resubmit"(修正後再提出)または"Reject"(掲載不可)のいずれかを決定します．"Reject"の場合には，この先には進めません．
4	2〜5カ月	論文修正 (Revision)	"Revise and resubmit"の場合には，査読者のコメントを参考にして，論文を修正しましょう．ほとんどの場合，査読者のコメントは，論文を改善するのに役立ちます．
5	2〜5カ月	修正論文投稿 (Submission of the revised manuscript)	修正した論文と査読者コメントに対する回答書を，論文投稿用ウェブページから電子投稿します．回答書は，下記のように査読者の各コメントと回答を対にし，また論文中の対応する修正部分のページ番号を記し，さらに論文中の修正部分を下線などで目立たせておくと，修正論文の査読も効率よく行われるはずです． Authors' Responses to Reviewers' Comments We thank the reviewer's comments that helped us improve the paper. (1) Reviewer's remarks to the authors: It would be useful to specify the range of validity of Equation (5). Authors' Responses: The range of validity of Equation (5) has been added to the end of Section 2.2 (see page 4).
6	3〜7カ月	再投稿論文の査読結果通知 (Notification of	編集長から再投稿論文の査読結果と査読者のコメントを知らせる電子メールが届きます．編集長は，査読結果に基づき，"Accept"，"Revise and resubmit"または"Reject"のいずれ

		decision on the revised manuscript)	かを決定します．"Reject" の場合には，この先には進めません．"Revise and resubmit" の場合には，論文を修正し，再々投稿します．
7	3〜7カ月	最終原稿アップロード (Uploading the final files of the paper text and figures)	"Accept" の場合には，論文テキストおよび図表の最終原稿を，アップロードするようにとの指示が編集長からあります．指定されたファイル形式のテキストおよび図表を作成し，アップロードしましょう．著作権譲渡書も送るようにと指示があります．電子的に送れるものもあれば，郵送しなければならない場合もあります．
8	4〜9カ月	校正 (Proofreading)	校正刷り(Proof)が送られてきます．誤字や脱字などがないことを確認するため，真剣に読みましょう．訂正点は，できるだけ早く担当者に連絡しましょう(付4.3に電子メールの文例を示しています)．掲載料支払いの連絡もあります．必要事項を記入して，できるだけ早く送りましょう．
9	5〜12カ月	掲載 (Published)	いよいよ掲載です．雑誌に掲載された論文をもう一度読んで，喜びを噛み締めましょう．

付 1.2 国際会議での論文発表までの流れ
General Flow of Procedures Needed for an Oral Presentation at an International Conference

本付録では，国際会議の論文募集要項を入手してから論文発表にいたるまでに必要な手続きの流れを示します．**付表 1.2** は，国際会議としておそらく最も一般的な 2 年ごとに開催されるもの(biennial conferences)を例として，必要手続き，実施時期および関連事項をまとめたものです．なお，会議を実質的に表す英語としては，"conference" 以外に，"assembly"，"colloquium"，"convention"，"meeting"，"symposium" などがありますが，それらの相違を気にする必要はないと思います．

付表 1.2 2 年ごとに開催される国際会議での論文発表に必要な手続きの流れ

順序	およその時期	必要手続き，国際会議事務局などからの連絡	説　　　明
1	12〜18カ月前	論文募集要項入手 (Call for papers)	国際会議のウェブページから入手するのが一般的です．論文募集分野，開催日程，投稿期日，論文フォーマットなどの情報を得ましょう．
2	10カ月前	論文要旨投稿 (Submission of an abstract)	指定されたフォーマットで作成した論文要旨を電子投稿しましょう．
3	8カ月前	論文仮採択または不採択通知	論文の仮採択または不採択通知が電子メールで送られてきます．不採択の場合には，この先には進めません．この段階で

		(Notification of provisional acceptance or rejection)	不採択となるのは，投稿した論文要旨の内容が会議の論文募集分野と著しく異なる場合などです．
4	6カ月前	論文投稿 (Submission of a full paper)	指定されたフォーマットおよびページ数(4または6ページの場合が多い)で作成した論文を電子投稿しましょう．
5	4カ月前	論文採択または不採択通知 (Notification of final decision on the paper)	論文の採択または不採択通知が電子メールで送られてきます．採択または不採択の結果のみを知らされる場合もあれば，論文に対する査読者のコメントも含めて知らされる場合もあります． パスポートをまだ持っていない場合には，この時点(あるいはそれ以前に)で，申請手続きを開始しましょう．また，開催国に入国するのにビザ(査証)が必要な場合にも，この時点で申請手続きを開始しましょう．国際会議のセクレタリーに，ビザ申請に必要な招待状を送っていただきましょう(付4.6に電子メールの文例を示しています)．
6	3カ月前	論文最終版投稿 (Submission of the final paper)	採択が決まっている場合には，論文中に誤りや見にくい部分が含まれていないことを確認して，最終版を電子投稿します．査読者のコメントが知らされている場合には，それらを参考にして論文を改善してから投稿します．
7	1カ月前	会議参加事前登録 (Advanced registration)	会議への参加費を，クレジットカードなどで事前に支払うのが一般的になっています．この時期までには，国際会議のウェブページにプログラム(予定表)が掲載されていると思います．自分の論文の発表日や時間を確認しておきましょう．また，滞在予定期間中のホテルや航空券も，採択が決まったら予約しておきましょう．
8	国際会議開催期間中	論文発表 (Presentation of the paper)	セッションが始まる前の休憩時間に，USBメモリスティックなどにコピーして持参した発表用のスライドファイルを，会場のパソコンにコピーし，正しく表示されることを確認しておきましょう．事前に，ウェブディスクにスライドファイルをアップロードするように指示される場合もあります．

※ 順序2，3の過程がない国際会議も多く存在します．

付　　録

付録2　インパクトファクタについて
On Journal Impact Factors

　学術雑誌のインパクトファクタ（impact factor）という用語を耳にしたことはありませんか．インパクトファクタとは，ある雑誌に掲載された1論文あたりの年間平均被引用回数で，雑誌の影響力を示す指標の一つです．この指標が高いほど，その雑誌は影響力の高い論文を収録しているといえます．このことから，同じ分野の複数の雑誌のなかからインパクトファクタの高い雑誌を選んで論文を投稿することは，より多くの専門家の目に触れる可能性を高めるため，意味があると考えられています．

　インパクトファクタは，毎年クラリベイト（Clarivate）社の引用文献データベースに収録されるデータをもとに次のように算出されています（https://clarivate.com/webofsciencegroup/essays/impact-factor/）．例えば，ある雑誌が，2020年に290編の論文を，2021年に310編の論文を掲載し，2022年に前者が830回，後者が970回引用されたとすると，その雑誌の2022年におけるインパクトファクタは$(830+970)/(290+310)=3.00$になります．この計算法からも明らかなように，インパクトファクタは，その雑誌が過去2年間に掲載した論文の年間平均被引用回数を示す指標です．当然，その雑誌には，数十回以上引用される論文も掲載されていれば，ほとんど引用されない論文も掲載されています．したがって，論文の価値については，雑誌のインパクトファクタではなく，それぞれの論文の総被引用回数で評価するのがより適切です．

　ただし，クラリベイト社のデータベースにすべての雑誌が収録されているわけではありません．当然，収録対象外の雑誌に掲載された論文の被引用回数は数えられませんし，その雑誌のインパクトファクタも算出されません．このため，雑誌が，このデータベースの収録対象になることは，よい論文の投稿先となる条件の一つといえます．

　アメリカ電気電子学会（IEEE: The Institute of Electrical and Electronics Engineers）やイギリス工学技術学会（IET: The Institution of Engineering and Technology）の論文誌は，多くのよい論文を収録しています．しかし，以前は，比較的査読期間が長く，さらに掲載が決定してから実際に掲載されるまでの期間も長いという欠点がありました．また，これらは電気電子系の主要学会誌であるため，これらの雑誌に掲載された論文は，これらの雑誌で引用される可能性も高くなります．このこと自体は悪くはないのですが，掲

載にいたるまでの期間が長いことと相まってしまうと，本来の質や影響力に対応していない低いインパクトファクタを得てしまうことになります．現在では，この状況はかなり改善されており，例えば，IEEE Transactions on Power Delivery のインパクトファクタは 1.0 未満であった時期もありましたが，2022 年のインパクトファクタは 4.4 にも至っています．2006 年 5 月に発刊された電気学会(IEEJ: The Institute of Electrical Engineers of Japan)の共通英文誌(IEEJ Transactions on Electrical and Electronic Engineering)のインパクトファクタは，2010 年には 0.36 でしたが，2022 年には 1.0 に上昇しています．

付録

付録3 h 指数について
On the h-index

　最近は，研究者の学術分野での貢献度を 1 つの数値で示す h 指数(h-index)(Hirsch 2005, Proceedings of the National Academy of Sciences of the United States of America)と呼ばれる指標が使用されるようになりました．インパクトファクタの高い著名な雑誌に掲載された論文が多数あっても，それら全ての被引用数が多いとは限らないため，論文の総発表数や著名雑誌への総発表数のみでは，その研究者の貢献度を示すことは困難でした．そこで，研究者の発表論文数と被引用数の両方に基づいた貢献度を表すために考案されたのが h 指数です．この指標は次のように算出されます．ある研究者の h 指数は，その研究者が発表した論文のうち，被引用数が h 以上である論文数が h 以上あることを満たすような最大の数値です．具体的には，h 指数が 20 である研究者は，被引用数 20 以上の論文を少なくとも 20 編発表していることを示しています．この指標には，研究者の論文の数(発表数)と質(被引用数)を 1 つの数値で表すことができるという利点があります．

　例えば，学術論文等の検索サービスである Google Scholar に登録すると，自身の h 指数を知ることができます．Google Scholar に登録すると，業績が自動的に更新・管理され，h 指数だけではなく，自身の論文の総被引用数，各年の総被引用数，各論文の被引用数や引用先の論文情報(題目，著者，発表年，雑誌名など)なども知ることができます．エルゼビア社の Scopus やクラリベイト社の Web of Science も同様のサービスを提供していますが，これらはインパクトファクタが付与されている雑誌に掲載された論文が主対象となっているようです．一方，Google Scholar は，日本語論文やインパクトファクタの付与対象となっていない雑誌に掲載された論文も対象にしていますので，Google Scholar に基づく h 指数の値は Scopus や Web of Science に基づく値より高くなります．

付録 4　電子メール文例
Examples of Electronic Correspondence (email)

本付録では，英語論文の執筆時や執筆後に役立つ可能性のある電子メールの文例を紹介します．

付 4.1　論文査読状況の照会
Inquiry About the Current Status of a Submitted Paper

投稿した論文の査読結果が，投稿後 1 カ月以内に得られる場合もあれば，半年以上経過してもなんの応答もない場合もあります．後者の場合に，編集長あるいは担当の編集委員に送る状況確認の電子メールの文例を以下に示します．

Subject　：　Inquiry about the current status of our paper
　　　　　　（Manuscript ID: TEMC-XXX-2021）
　　　　　　（論文　TEMC-XXX-2021 の現状についての問い合わせ）

Dear Prof. Jane Doe（編集長または担当編集委員の名前），

Could you please let me know the current status of the following paper, which I submitted about six months ago?
（半年ほど前に投稿いたしました下記論文の現状をお知らせいただけませんでしょうか．）

Manuscript ID　：　TEMC-XXX-2021
Authors　　　　：　Hanako Denki, and Taro Denshi
Paper title　　　：　Voltages induced on an overhead wire by lightning strikes

Sincerely,
Hanako

```
Hanako Denki
Ph.D. student
University of Southern Kyoto
Department of Electrical and Electronics Engineering
Kyotanabe, Kyoto ○○○-△△△△, Japan
Phone  :   +81-□□□-○○-△△△△
Fax.   :   +81-□□□-○○-△△□□
E-mail :   ○○○○@△△△△.ac.jp
```

〈補足〉 "Sincerely" は手紙や電子メールの一般的な締めの挨拶です．"Best regards", "Kind regards", "Regards" などもよく用いられています．

付4.2 著作権で保護された図の使用許可願い
Request of Permission to Reprint Copyrighted Figures

ほかの学会や出版社の雑誌に掲載されている図表などを引用する場合には，その学会あるいは出版社から使用許可を得ておく必要があります．多くの場合，対応する学会や出版社の知的財産権オフィス(Intellectual Property Rights Office)に下記のような電子メールを送れば，使用許可を示す電子メールまたは使用許可を得るために必要な費用などを示す電子メールが返信されてきます．

Subject : Request of permission to reprint IEEE copyrighted figures
〔アメリカ電気電子学会(IEEE)が著作権を有する図の使用許可願い〕

Dear Sir or Madam,

This is to request permission to reprint two IEEE copyrighted figures. The figures that I would like permission to reprint are Figures 4 and 5 of the following paper:
(アメリカ電気電子学会(IEEE)が著作権を有する二つの図の使用を許可してください．使用を許可していただきたいのは下記論文の図4および図5です．)

Author	:	Jane Doe
Paper title	:	Electromagnetic models of the lightning return stroke
Journal title	:	IEEE Transactions on Electromagnetic Compatibility
Volume and issue numbers	:	vol. 1020, no. 5
Publication year	:	2020

I am going to reprint these figures in my paper that will be published in an upcoming issue of the following journal:
(これらの図を，下記雑誌の近刊号に掲載予定の論文で使用する予定です．)

Authors	:	Hanako Denki, and Taro Denshi
Paper title	:	Lightning electromagnetic environment
Journal title	:	Journal of Geophysical Research
Publisher's name	:	American Geophysical Union

The citation to the figures is as follows:
J. Doe, "Electromagnetic models of the lightning return stroke," IEEE Trans. EMC, vol. 1020, no. 5, 2020.

I would appreciate (it) if you would reply to this email and let me know if there is any applicable fee.
(このメールにご返信いただき，これらの図の使用料が必要か否かについて，お知らせいただければありがたく存じます．)

Thank you in advance.
(よろしくお願いいたします．)

Sincerely,
Hanako

Hanako Denki
Ph.D. student

付　　　録

```
University of Southern Kyoto
Department of Electrical and Electronics Engineering
Kyotanabe, Kyoto ○○○-△△△△, Japan
Phone    :    +81-□□□-○○-△△△△△
Fax.     :    +81-□□□-○○-△△△□□
E-mail   :    ○○○○ @ △△△△.ac.jp
```

〈補足〉 "Dear Sir or Madam" はメールを受け取る方の名前も性別もわからない場合によく使われる敬辞です．"To whom it may concern"（関係各位）という表現もあります．"I would appreciate if you would（または could）…"，"Thank you in advance." もよく使う表現です．

　なお，最近では，出版元の論文のウェブページに，"Request permission for reuse" という，Copyright Clearance Center のウェブページへのリンクボタンが設けられており，雑誌論文や著書に掲載されている図やテキストの著作権許諾を（有料で）即座に得ることができるようになってきています．

付 4.3　論文校正刷りの訂正依頼
Request to Correct Errors in a Proof

　査読を経た論文が受理されてしばらくすると，校正刷り（proof）が送られてきます．校正刷りに誤字（typographical error または typo）や脱字（omission）などが見つかった場合には，下記のような訂正依頼のメールを編集者に送りましょう．

Subject　:　Request to correct errors in the proof (Manuscript ID: TEMC-XXX-2021)
　　　　　　（論文　TEMC-XXX-2021 の校正刷りの訂正依頼）

Dear ○○○○〔校正刷りの送信者（編集者）の名前〕,

Thank you very much for preparing the proof of our paper. Please incorporate the following three corrections in the final version of this paper.
（私たちの論文の校正刷りを作成してくださいましてありがとうございました．この論文の最終版には，下記の三つの訂正を加えてください．）

1) Please delete "channel" located between "lightning return-stroke" and "current" on line 100 on page 2.
   (2 頁 100 行目の "lightning return-stroke" と "current" の間の "channel" を削除してください．)
2) Please replace "sigma" on line 200 on page 4 by its corresponding Greek letter.
   〔4 頁 200 行目の "sigma" をギリシャ文字(σ)に置き換えてください．〕
3) Please insert "return-stroke " between "lightning" and "current" on line 300 on page 6.
   (6 頁 300 行目の "lightning" と "current" の間に "return-stroke" を挿入してください．)

Sincerely,
Hanako

Hanako Denki
Ph.D. student
University of Southern Kyoto
Department of Electrical and Electronics Engineering
Kyotanabe, Kyoto ○○○-△△△△, Japan
Phone ： + 81-□□□-○○-△△△△
Fax. ： + 81-□□□-○○-△△□□
E-mail ： ○○○○ @ △△△△.ac.jp

付 4.4　雑誌バックナンバーの注文　*Order of a Back Issue*

　学会の出版販売関連部門に下記のような電子メールを送れば，雑誌のバックナンバーを注文できます．多くの場合，注文用紙も添付してくれますので，購入に使うクレジットカードの情報を記入し，署名をして，ファクスなどで送り返せばよいでしょう．なお，たとえば，IEEE が出版する雑誌群のなかでよく参照する雑誌がある場合には，IEEE の会員になり，その雑誌の電子版を定期購読すると，その雑誌に過去に掲載された論文も参照できます．

付　　　録

Subject　：　Order of the issue (vol. 1020, no. 5) of IEEE Trans. EMC 　　　　　　（IEEE Trans. EMC の第 1020 巻 5 号の注文）

Dear Sir or Madam,

I would like to purchase a printed version of the following issue of the IEEE Transactions on Electromagnetic Compatibility.
（IEEE Trans. EMC の下記のバックナンバーを 1 冊購入したいと考えております．）

Journal title　　　　　　　　：　IEEE Transactions on Electromagnetic Compatibility
Volume and issue numbers：　vol. 1020, no. 5
Publication year　　　　　　：　2020

Please let me know the price of this issue and its air carriage to Japan.
（このバックナンバーの価格と日本への航空郵送料を教えてください．）

Sincerely,
Hanako

Hanako Denki
Ph.D. student
University of Southern Kyoto
Department of Electrical and Electronics Engineering
Kyotanabe, Kyoto ○○○-△△△△, Japan
Phone　：　＋81-□□□-○○-△△△△
Fax.　　：　＋81-□□□-○○-△△□□
E-mail　：　○○○○ @ △△△△.ac.jp

付 4.5　論文送付願い　*Request to Send a Copy of a Paper*

最近では，学会や出版社の論文データベースから雑誌に掲載された論文や著名な国際会議で発表された論文の電子ファイルを（多くの場合，有償で）入

手できることが多くなっています．しかし，あまり有名ではない国際会議で発表された論文については，入手が困難な場合があります．このような場合，下記のような電子メールを著者に送れば，その論文の電子ファイルを送ってくれる場合があります．

Subject ： Request to send your paper presented at ICLP 2010
（国際会議 ICLP 2010 で発表された論文の送付願い）

Dear Prof. Jane Doe,

My name is Hanako Denki, a Ph.D. student of the University of Southern Kyoto. I have read your papers published in journals with great interest. I would like to read the following conference paper, which is found in the reference list of one of your journal papers. Could you please send me an electronic version of the paper, preferably in pdf format?
（電気花子と申します．南京都大学の大学院生です．雑誌に掲載されたあなたの論文を興味深く読ませていただきました．それらのなかの一つの参考文献リストにある下記の国際会議発表論文を読ませていただきたいと考えております．その論文の電子ファイルを，できれば pdf 形式のものを私にお送りくださいませんでしょうか．）

Author ： Jane Doe
Paper title ： Lightning electromagnetic field computation
Conference name ： International Conference on Lightning Protection
Year ： 2010

Thank you in advance.
（よろしくお願いいたします．）

Sincerely,
Hanako

Hanako Denki
Ph.D. student
University of Southern Kyoto
Department of Electrical and Electronics Engineering

付　　　録

```
Kyotanabe, Kyoto ○○○-△△△△, Japan
Phone  :   +81-□□□-○○-△△△△
Fax.   :   +81-□□□-○○-△△□□
E-mail :   ○○○○@△△△△.ac.jp
```

付 4.6　ビザ（査証）取得のための招待状送付願い
Request of an Invitation Letter for Acquiring a Visa

　国によっては，入国するのにビザ（査証）が必要な場合があります．そのような国で開催される国際会議に参加するためには，その国際会議のセクレタリーに，下記のような電子メールを送れば，ビザ申請に必要な招待状を送ってくれます．

Subject　：　Request of an invitation letter for acquiring my visa for India
（インド入国ビザ取得のための招待状送付願い）

Dear Sir or Madam,

My name is Hanako Denki. I am an author of the following paper, which has been accepted for presentation at the International Conference on Lightning (ICL 2022) in New Delhi, India.
（電気花子と申します．私は，インドのニューデリーで開催される雷に関する国際会議 ICL2022 での発表が決定しております下記論文の著者です．）

Paper number：　ICL2022-E001
Authors　　　：　Hanako Denki, and Taro Denshi
Paper title　　：　Electromagnetic computation method for lightning protection studies

Could you please send an invitation letter for attending this conference to me? I need it to acquire my visa for India.
（ビザ取得に必要ですので，この会議への招待状をお送りくださいませんでしょうか．）

Sincerely,
Hanako

Hanako Denki
Ph.D. student
University of Southern Kyoto
Department of Electrical and Electronics Engineering
Kyotanabe, Kyoto ○○○-△△△△, Japan
Phone ： ＋81-□□□-○○-△△△△
Fax. ： ＋81-□□□-○○-△△□□
E-mail ： ○○○○＠△△△△.ac.jp

付 4.7　博士研究員としての留学の打診
Application for a Postdoctoral Position

国際的な環境で研究を行うことは，英語力を伸ばせるだけではなく，新しい思考法も身につき，とても価値があります．ここでは，国内の大学院で博士後期課程を終了後，博士研究員としての留学の打診をする場合の電子メールの文例を示します．

Subject ： Application for a postdoctoral position
（博士研究員ポストへの応募）

Dear Prof. Jane Doe,

My name is Hanako Denki, a Ph.D. student in Electromagnetics Laboratory at the University of Southern Kyoto, Japan. I am writing to inquire about the possibility of a postdoctoral position in your Lightning Laboratory.
（電気花子と申します．日本の京都にある南京都大学電気磁気学研究室の博士課程の学生です．ジェーン教授の雷研究室の博士研究員ポストについてお伺いしたいことがございます．）

My Ph.D. work has focused on developing a new method for computing lightning electromagnetic fields. This computation method is based on a

numerical solution of Maxwell's equations in the time domain. I have tested its validity by comparing computed results with corresponding measured results, and found it to be sufficient. The fundamental part of this work has already been published in the IEEE Transactions on Electromagnetic Compatibility, and an advanced part of it has been accepted for publication in the same journal.
〔博士課程において，私は雷電磁界の新しい計算法の開発に重点的に取り組んできました．この計算法は，マクスウェル方程式の時間領域解法に基づいています．この計算法による結果と対応する実測結果との比較を行うことにより，この計算法の妥当性をテストし，十分妥当であることがわかりました．この研究の基本部分は，すでにアメリカ電気電子学会電磁環境部門誌（IEEE Trans. Electromagnetic Compatibility）に掲載されており，それより一歩進めた内容の論文も同じ雑誌に掲載されることが決まっています．〕

I have recently become interested in modeling of lightning leader and return-stroke processes. I think that my electromagnetic computation method could be of great use in performing this research. Thus, I am quite interested in working with you in your Lightning Laboratory.
（最近，私は雷の前駆放電と帰還雷撃過程のモデリングに興味をもっております．私の電磁界計算法がこの研究を行うのに非常に役立つのではないかと思っております．このように，私はジューン先生の雷研究室で研究を行うことに大変興味をもっております．）

I will defend my Ph.D. most likely by March 2022. I would like to start my postdoctoral work with you between April and September 2022.
（私は，2022年の3月に博士号を取得予定です．2022年の4月から9月までの間に，ポスドクとして研究を開始したいと考えております．）

Attached is my C.V., and I look forward to hearing from you.
（履歴書を添付しております．お返事をいただければ幸いです．）

Sincerely,
Hanako

Hanako Denki

```
Ph.D. student
University of Southern Kyoto
Department of Electrical and Electronics Engineering
Kyotanabe, Kyoto ○○○-△△△△, Japan
Phone  ：  ＋81-□□□-○○-△△△△
Fax.   ：  ＋81-□□□-○○-△△□□
E-mail ：  ○○○○＠△△△△.ac.jp
```

〈補足〉 "I am writing to..." はよく使う表現です．"Attached is (または are)..." もよく使う表現です．封書の場合は，"Enclosed is (または are)..."（... を同封致します）とします．

```
                                                    Nov. 1, 2021
                       Curriculum Vitae
                          （履歴書）

                        Hanako Denki

Hanako Denki
Ph.D. student
University of Southern Kyoto
Department of Electrical and Electronics Engineering
Kyotanabe, Kyoto ○○○-△△△△, Japan
Phone  ：  ＋81-□□□-○○-△△△△
Fax.   ：  ＋81-□□□-○○-△△□□
E-mail ：  ○○○○＠△△△△.ac.jp


Objective
（目的）
Postdoctoral position in electromagnetics
（電気磁気学における博士研究員ポスト）


Personal
（個人情報）
First name    ：   Hanako
```

付　　録

Last name　　：　Denki
Home address：　Uji, Kyoto △△△-□□□□, Japan

Education
(学歴)
March 2013　　Kyoto Central High School
March 2017　　B.Sc. degree (in Electrical Engineering)
　　　　　　　University of Southern Kyoto, Japan
　　　　　　　Title of graduation thesis: Equivalent-circuit modeling of grounding electrodes for surge simulations (in Japanese)
March 2019　　M.Sc. degree (in Electrical Engineering)
　　　　　　　University of Southern Kyoto, Japan
　　　　　　　Title of master's thesis: Application of the method of moments to analyzing lightning surges
March 2022　　Ph.D. (in Electrical Engineering) *anticipated* (取得見込み)
　　　　　　　University of Southern Kyoto, Japan
　　　　　　　Title of doctoral thesis: Development of an electromagnetic computation method for lightning electromagnetic fields and surges

Employment
(職歴)
April 2019-present　　Research Fellow of the Japan Society for the Promotion of Science
(2019 年 4 月～現在　　日本学術振興会特別研究員)

Award
(受賞歴)
March 2019　　Outstanding Master's Thesis Award from the University of Southern Kyoto
(2019 年 3 月　　南京都大学優秀修士論文賞)

Language Skills
(語学力)

English (TOEFL score 620)
Japanese (native language)

Areas of Interest
(専門分野)
Lightning, Computational Electromagnetics, Electromagnetic Compatibility

Publications
(発表文献)
[1] H. Denki, and T. Denshi, "An electromagnetic computation method for lightning electromagnetic fields," IEEE Trans. Electromagnetic Compatibility, vol. 1021, no. 5, pp. 500-509, Nov. 2021.
[2] H. Denki, and T. Denshi, "Lightning electromagnetic environment in the presence of a tall grounded object," IEEE Trans. Electromagnetic Compatibility (in press) (印刷中).

References
(信用照会先)
Prof. Taro Denshi, University of Southern Kyoto, Japan
Phone　：　+81-□□□-○○-△△××
E-mail　：　○○××@△△△△.ac.jp
Prof. Jiro Joho, University of Tokyo, Japan
Phone　：　+81-□-□□○○-△×××
E-mail　：　△△××@□□□.ac.jp

〈補足〉 履歴書は "Curriculum Vitae" (C.V.) または "Résumé" といいます. 理学のみではなく, 理系の学士号および修士号は "Bachelor of Science degree" および "Master of Science degree" といい, それぞれ "B.Sc. degree" または "B.S. degree", "M.Sc. degree" または "M.S. degree" と略記されます. 日本語との対応を考えると少し奇異ですが, 農学も工学も応用科学 (Applied Science) の1分野ということだと思います. 日本語の修士(工学) または工学修士の「工学」をどうしても示したい場合には, "M.Sc. degree in Engineering" とするのがよいようです. より詳細に, "M.Sc. degree in Electrical Engineering" と記す場合もあるようです. "Master of Engineering" とはあまり記さないようです.

付　録

　文系理系に関係なく，博士号は"Ph.D."と記されます．これは，ラテン語の"Philosophiae Doctor"（英語の"Doctor of Philosophy"）の略記で，直訳すると哲学博士になりますので，修士（工学）と"M.Sc"の対応よりもはるかに奇異に思えます．19世紀前半ごろまでは，現在の自然科学（Natural Science）に対応する学問分野を示す用語として自然哲学（Natural Philosophy）が用いられていたようで，"Philosophy"という語はかなり広い範囲の学問一般を意味していたことになります．なお，イギリスには，Ph.D.より上級の学位として，"Doctor of Science"（DSc）があります．また，ヨーロッパには，博士論文の指導をするためには，"Habilitation"と呼ばれる資格（教授資格）が必要な国もあるようです．

　学位論文は"thesis"といいます．博士論文は，"Ph.D. dissertation"という場合もあります．

付録5　アメリカ留学関連手続き
Procedures Needed for Studying in the United States

付5.1　日本出国前の準備
Preparation in Japan Before Leaving for the United States

　ここでは，アメリカ留学前に国内ですませておかなければならない，あるいはすませておきたい手続きについて説明します．ただし，ここでの説明は，留学受入れ先が決まっており，そこでの身分が客員研究員または客員教員（正規の所属は日本の大学，研究所，企業など）であることを想定しています．今後，手続きの方法や提出書類が変わる可能性がありますし，現状でも，受入れ先や，そこでの身分によって異なる可能性もありますので，参考資料としてご覧になってください．

付表5.1　日本出国前の手続き

手続き	説　　明
滞在予定地の情報収集	アメリカの旅行・生活ガイドブックや，滞在予定州あるいは近隣州にある日本総領事館のウェブページを見て，滞在予定地の生活や安全に関する基本情報を得ておきましょう．また，留学予定の大学のウェブページで，その大学や研究室についての情報も得ておきましょう．
留学受入れ先発行のプログラム参加資格証明書（DS-2019）取得	DS-2019は，留学予定先が一定期間にわたって受け入れることを保証する証明書で，これがなければ交流訪問者ビザ（J-1ビザ）を申請できません．この取得には，申請後1〜2カ月程度必要な場合が多いので，留学が決まったら，できるだけ早く必要な書類をそろえて申請しましょう． DS-2019の申請には，下記の書類を留学受入れ先に提出する必要があります． ・DS-2019申請書（受入れ先から入手．オンライン申請できる大学も増加） ・パスポートのコピー ・現在の勤務先の上司からの英文推薦状 ・英文の履歴書 ・英文の卒業証明書，修了証明書 ・英文の財政証明書（滞在費用を賄える十分な資金があることを証明するもの：銀行残高証明書，日本の派遣元からの資金提供を示す書類など）
ビザ取得	アメリカに91日以上滞在するためにはビザが必要です．客員の教員，研究員，大学院生としてアメリカの大学に招かれる場合に取得しなければならないビザの種類は，交流訪問者ビザ（J-1ビザ）です． 同行家族は，交流訪問者同行家族ビザ（J-2ビザ）を取得しておく必要があります．

	J-1 ビザの申請には，下記の書類を駐日アメリカ大使館または総領事館に提出する必要があります． • パスポート • DS-2019 • オンラインビザ申請書(DS-160)(アメリカ大使館ウェブページの該当ページでオンライン入力．証明写真のアップロード必要．ビザ申請料支払必要) • 面接予約(アメリカ大使館ウェブで DS-160 の申請後にオンライン予約) • 英文の最終卒業学校成績証明書 • 英文の最終卒業学校の卒業証明書 • I-901 SEVIS 管理費支払い証明書 • 英文の財政証明書(滞在費用を賄える十分な資金があることを証明するもの：銀行残高証明書，日本の派遣元からの資金提供を示す書類など) • 返信宛先を記入した返信用封筒(レターパックプラス) J-2 ビザの申請にも，J-1 ビザの申請と同じ書類が必要ですが，J-1 ビザと同時に申請する場合には，上記のうち，成績証明書，I-901 SEVIS 管理費支払証明，財政証明書および返信用封筒は不要です． アメリカの大学あるいは大学院の正規学生として留学する場合，J-1 ビザではなく，留学生用ビザ(F-1 ビザ)を取得する必要があります．この場合，同行家族は F-2 ビザを取得しなければなりません．上述した J-1，J-2 ビザ申請と異なるのは，DS-2019 ではなく I-20(留学先大学または大学院への入学許可証)の提出が必要になる点です．
海外旅行保険加入	アメリカで病気や怪我の治療を受けると，高額な費用がかかります．留学受入れ先の大学は，滞在期間中の保険加入を義務づけていると思いますので，その条件を満たすものに加入しておきましょう．また条件を明記した英文の保険加入者証も得ておきましょう．出国前に健康診断を受け，また病気や歯の治療を受けておきましょう．
国際クレジットカード加入	アメリカでは，ほとんどすべての支払い時にクレジットカードが使用できますので，加入しておきましょう．使用した金額は日本の銀行口座から月々決済されますので，使用明細をインターネットで確認できるようにしておくと便利です．スマートフォンによる決済は，日本ほどはまだ普及していないようです．
国際運転免許証取得	都道府県の運転免許試験場で発行してもらえます．この免許証の有効期限は交付されてから 1 年ですが，アメリカに居住する場合には，住居が決まってから定められた期間内に居住州の運転免許証を取得しなければなりません(期間は州によって異なります．たとえばフロリダ州の場合には 30 日以内です)．
航空券の予約と手荷物の選定 (付表 4.2 参照)	渡米のための航空券を予約しておきましょう．アメリカ発着便の無料手荷物許容量は，航空会社および座席クラスによって異なります．航空券購入前に，無料手荷物許容量および超過料金などを調べておきましょう．たとえば，ある日系航空会社のエコノミークラスを利用する場合の無料手荷物許容量は，荷物の個数が 2 個以内，それぞれの荷物の 3 辺の和が 203 cm 以内，かつ，それぞれの重量が 23 kg 以内となっています．
荷物の送付	手荷物以上の荷物がある場合には，郵便局や引越し業者を利用して発送する必要があります．荷物が少なければ，航空便を利用するとよいでしょう．船便の場合には，到着するのに 1, 2 カ月かかるので，ある程度前に送っておく必要があります．ただし，通関用の書類にはビザについて記載する欄があるので，この時期までにはビザを取得しておかなければなりません． 本人が現地に到着して引越し先が決まったら，引越し業者のアメリカの連絡先に電

	話をして，配達を依頼しましょう．
仮住まいの予約	アメリカでの住居が決まるまでの間は，ホテルなどに滞在しなければなりません．大学に近いホテルの部屋を数日間分予約しておきましょう．
米ドルへの両替	アメリカ到着後，数日間過ごせる程度の米ドルを現金で準備しておきましょう．

付表5.2　手荷物として持参したいアイテム

アイテム	説　明
英会話本	アメリカで生活すれば，英語は徐々に上手になっていきます．しかし，現実には，入国直後に，住居，銀行，電話，水道，電気などの契約や大学での諸手続きを英語で行わなければなりません．そのためにも，役立つ表現を多く載せた英会話本を持参すると便利です．
ノートパソコン	日本語OS，日本語ソフトをインストールした使い慣れたものを持参しましょう．なお，日本の電源電圧は100 V（実効値），周波数は50または60 Hzですが，アメリカの電源電圧は120 V（実効値），周波数は60 Hzです．持参しようと考えている電気製品がある場合には，その説明書をみて，電圧（実効値）120 V，周波数60 Hzでも使用可能か否かを確認しておきましょう．ノートパソコンは，通常付属しているACアダプタにより，電源側電圧100～240 V，周波数50または60 Hzの交流を，たとえば19 Vの直流に変換しているため，そのまま使用できることがほとんどだと思います．電源コンセント（wall outlet）の形は日本のものと同じA-2タイプです．
電子辞書	契約書や説明書を読む際に，とても役立ちます．スマートフォンやノートパソコンのオンライン辞書やオンライン翻訳を使い慣れている場合には，それでもよいと思います．なお，電子機器用の乾電池については，日本のものと同じ電圧（1.5 V）で同じサイズのものがアメリカの電器店やドラッグストアで手に入ります．ただし，呼び方は違っていて，単1がD，単2がC，単3がAA（double A battery），単4がAAA（triple A battery），単5がNと呼ばれています．ボタン形およびコイン形電池についても，ほとんどのものがアメリカで購入できます．これらの電池については，日本では国際電気標準会議（IEC: International Electrotechnical Commission）の定める名称で呼ばれていますが，アメリカでは異なっています．たとえば，日本のボタン形電池LR44は，ENERGIZER社のA76やDURACELL社のPX76Aに対応しており，コイン形電池CR2032は，ENERGIZER社のECR2032やDURACELL社のDL2032に対応しています．

付5.2 アメリカ入国後の手続き
Procedures Needed Right After Arrival in the United States

ここでは,アメリカ入国直後にしなければならない諸手続きについて説明します.今後,手続きの方法は変る可能性がありますので,参考資料としてご覧になってください.

付表5.3 アメリカ入国後の手続き

手続き	説明
住居探し	安全な地域をホストの先生や研究室の大学院生に聞いてから探しましょう.アパート紹介業者を利用して条件にあうものを紹介してもらってもよいし,直接,アパートの事務所を訪ねて,部屋を見せてもらってもよいと思います.気に入ったアパートが見つかれば,仮契約して,近くの銀行に向かいましょう.また,荷物を日本から送った場合には,利用した引越し業者のアメリカの連絡先に電話をかけて,配送を依頼しましょう.
銀行当座預金口座開設	アメリカでは,家賃や電気,水道,電話の料金,そのほかの支払いに,今でも小切手(personal check)を使用することがあるので,このための当座預金口座(checking account)を開設しましょう.日本の銀行からの送金にも,アメリカの銀行の口座が必要になります. 口座開設には,アメリカでの住所,パスポートと預け入れる現金が必要です.最低預金額(minimum balance)が確保されていれば,その口座管理量が無料になります.口座を開設すると,その場で,小切手帳がもらえます.数日後には,自分の名前と住所が印刷された小切手帳が郵送されてきます.預金通帳はありませんが,毎月明細書が送られてくるか,インターネットで使用明細を確認できます. 口座開設時に,通常はデビットカード(debit card)も作ります.これは,商品購入時に銀行などの預金口座から即時に引き落として支払うクレジットカードに似たカードのことです.使用時点で預金残高がなければ支払いできない点が,クレジットカードと異なっています.また,このカードでATM(automated teller machine)で現金を引き出したり(withdraw),入金する(deposit)などができます.
電話,インターネット・プロバイダ契約	最寄の電話局に行って申し込みましょう.近距離,長距離,国際電話,インターネット,それぞれについていくつかのプランが用意されているので,説明を聞き,自分にあったプランを選びましょう.その場で保証金(deposit)も支払いましょう. 月々の支払い方法として,クレジットカードかパーソナルチェックが選べます.後者の場合には,毎月,使用明細,支払い額,返信用封筒を含む封筒が送られてきます.支払い額をパーソナルチェックに記し,返信用封筒に入れて,電話会社に返送します.切手は,自分で購入してなければなりません.シール式の切手が便利です.
電気,水道契約	電気,水道などの公共サービスをユーティリティーズ(utilities)と呼びます.最寄の営業所に行って契約しましょう.その場で保証金(deposit)も支払いましょう. 月々の支払い方法として,クレジットカードかパーソナルチェックが選べます.
食器,家具,電気製品の購入	食器,家具,電気製品は,近くのTargetやWalmartなどの大型複合小売店(big-box store)などで購入できます.

付録5　アメリカ留学関連手続き

客員研究員登録, 身分証明書交付申請	大学の国際センターを訪ね, 客員研究員または客員教員の登録手続きをし, 大学の身分証明書（カード）を交付してもらいましょう. 持参するものは, パスポート（J-1ビザの貼付されたもの）, DS-2019, 医療保険加入証明書などです. 保険は, 大学が指定する条件を満たすものでなければなりません. 日本で加入する場合には, 条件を明記した英文の保険加入者証を得ておきましょう.
医療保険加入	日本で加入した海外旅行保険の補償額が, 大学の指定する条件を満たしていない場合には, 別途医療保険を購入しなければなりません. 大学の国際センター内に保険案内または販売の窓口があることが多いので, そちらで購入することもできると思います.
外務省への在留届	外務省のウェブページからオンライン在留届を提出しておきましょう. この届出をしておくと, 巻き込まれる可能性のある大きな事件, 事故, 災害などが発生した際に, 在米日本大使館または領事館から, その情報を電子メールなどで知らせてくれます.
社会保障番号（SSN）の取得	アメリカには, 戸籍や住民票はないので, 社会保障番号（SSN: Social Security Number）が, それらの役割を果たしているといわれています. 受入れ先から給与が出る場合には, SSNを取得しなければなりません. SSNの申請には, 下記の書類を社会保障事務所に提出する必要があります. • SSN申請書（SS-5） • パスポート • DS-2019
研究提案書・計画書提出	これは必要なものではありませんが, 限られた留学期間を有効に使うためにも, ホストの先生に提出しておくのがよいと思います.
自動車免許取得	日本で取得した国際免許証の有効期限は交付されてから1年ですが, アメリカに居住し, 自動車を運転する場合には, 住居が決まってから定められた期間内に居住州の運転免許証を取得しなければなりません（期間は州によって異なります）. 運転免許証を取得するためには, 運転免許事務所（Driver License Office）を訪ね, まずは筆記試験を受けなければなりません. 筆記試験は, 予約なしで受験できます. 州の陸運局（Department of Motor Vehicles）のウェブページから運転者の手引書（Driver's guide）を入手して, 交通ルールを頭に入れてから臨みましょう. 受験に必要な書類は, SSN（持っていれば）, パスポート（ビザが貼付されたもの）, DS-2019などです. 筆記試験の合否は, 受験後すぐに知らされます. 合格したら, 実技試験を予約しましょう. 実技試験には, 自分の車かレンタカーを持ち込まなければなりません. 実技試験の合否も, 試験後すぐに知らされます. 合格者には, 写真撮影後, 仮の免許証が交付されます. 正式な免許証は, 後日郵送されてきます.
自動車購入	新車の価格は, 日本と同程度です. 中古車を購入する場合には, ブルーブック（Kelly Blue Book）のウェブサイトなどで, その車の適正価格を確認しておきましょう. なお, 公共の安全な交通機関, 徒歩あるいは自転車で安全に通学できる場合には, 自動車を購入しなくても生活できる場合もあると思います. ナンバープレートのデザインは多くのものが用意されています. 帰国時に持ち帰ることもできるので, 気に入ったデザインのものを選びましょう.
自動車保険購入	自動車を購入したら, 自動車保険も購入しなければなりません. 免許取得直後ということで安全運転者としての実績がないため, 補償内容に見合わないような高額の保険料を支払う必要があります. 保険とは異なりますが, 日本のJAFに相当するAAA（American Automobile Association）の会員になっておくと, 多くのサービスが受けられます.

付　録

付録6　電気学会誌に掲載された英語論文執筆，発表法に関する文献リスト
List of IEEJ Journal Articles on Writing and Presenting Scientific Papers in English

　下記に示しますように，英語論文などの執筆法や執筆時の心得，英語での発表法や発表能力を高めるための試み，留学や国際会議参加の意義を記した有用な記事が，電気学会誌に掲載されています．電気学会の会員であれば，同学会のウェブページから下記の文献を入手できますので，本書と一緒に手元においておくと役立つと思います．

英語論文執筆法について
[1]　富山健，富山真知子：「電気・電子技術者の英語論文作法（その1）」，電気学会誌，vol. 112, no. 8, pp. 625-628（1992）
[2]　富山健，富山真知子：「電気・電子技術者の英語論文作法（その2）」，電気学会誌，vol. 112, no. 9, pp. 726-730（1992）
[3]　富山真知子，富山健：「電気・電子技術者の英語論文作法（その3）」，電気学会誌，vol. 112, no. 10, pp. 805-809（1992）
[4]　長谷良秀：「日本語・英語の技術専門書出版体験から―所感　技術者は英語といかに付き合うべきか―」，電気学会誌，vol. 128, no. 8, pp. 550-552（2008）

英語論文発表法について
[1]　富山真知子，富山健：「電気電子技術者の英語口頭発表作法（その1）」，電気学会誌，vol. 113, no. 8, pp. 678-681（1993）
[2]　富山真知子，富山健：「電気電子技術者の英語口頭発表作法（その2）」，電気学会誌，vol. 113, no. 10, pp. 853-856（1993）
[3]　富山真知子，富山健：「電気電子技術者の英語口頭発表作法（その3）」，電気学会誌，vol. 113, no. 11, pp. 955-958（1993）
[4]　川田昌武：「研究室から始める国際化への試み」，電気学会誌，vol. 129, no. 5, p. 316（2009）

海外留学，国際会議参加の意義について
[1]　吉川庄一：「海外留学を考える方へ」，電気学会誌，vol. 112, no. 3, pp. 192-195（1992）

［2］ 大西公平：「国際会議発表ノススメ」，電気学会誌，vol. 115, no. 6, pp. 370-373（1995）

［3］ 村谷拓郎：「海外で働こう」，電気学会誌，vol. 131, no. 2, p. 67（2011）

［4］ 野崎貴裕：「世界がみんなを待っている!!」，電気学会誌，vol. 142, no. 12, pp. 778-781（2022）

［5］ 渡邉優太郎，B. R. Moser：「マサチューセッツ工科大学（MIT）留学記―アメリカで感じたイノベーション―」，電気学会誌，vol. 143, no. 8, pp. 525-528（2023）

参考文献

Anderson, J. G., L. E. Zaffanella, G. W. Juette, M. Kawai, and J. R. Stevenson (1971): "Ultrahigh-voltage power transmission," Proceedings of the IEEE, vol. 59, no. 11, pp. 1548-1556.

Balanis, C. A. (2005): Antenna Theory: analysis and design, John Wiley & Sons, Inc.

Bouhafs, F., M. Mackay, and M. Merabti (2012): "Links to the future," IEEE Power & Energy Magazine, vol. 10, no. 1, pp. 24-32.

Bird, J. (2003): Electrical Circuit Theory and Technology, Newnes.

Bollen, M. H. J. (2001): "Voltage sags in three-phase systems," IEEE Power Engineering Review, vol. 21, no. 9, pp. 8-11.

Comerford, R. (2000): "The Internet," IEEE Spectrum, vol. 37, no. 1, pp. 40-44.

Denholm, P., W. Cole, and N. Blair (2023): "Moving beyond 4-hour Li-ion batteries: Challenges and opportunities for long (er) -duration energy storage," Golden, CO, National Renewable Energy Laboratory, no. NREL/TP-6A40-85878.

Gupta, M., C. Akiri, K. Aryal, E. Parker, and L. Praharaj (2023): "From ChatGPT to ThreatGPT: Impact of generative AI in cybersecurity and privacy," IEEE Access, vol. 11, pp. 80218-80245.

Halbouni, A., T. S. Gunawan, M. H. Habaebi, M. Halbouni, M. Kartiwi, and R. Ahmad (2022): "Machine learning and deep learning approaches for cybersecurity: A review," IEEE Access, vol. 10, pp. 19572-19585.

Hassenzahl, W. V. (1983): "Superconducting magnetic energy storage," Proceedings of the IEEE, vol. 71, no. 9, pp. 1089-1098.

Hirsch, J. E. (2005): "An index to quantify an individual's scientific research output," Proceedings of the National Academy of Sciences of the United States of America, vol. 102, no. 46, pp. 16569-16572.

Holttinen, H., A. G. Orths, P. B. Eriksen, J. Hidalgo, A. Estanqueiro, F. Groome, Y. Coughlan, H. Neumann, B. Lange, F. van Hulle, and I. Dudurych (2011): "Currents of change," IEEE Power & Energy Magazine, vol. 9, no. 6, pp. 47-59.

Li, Y., J. Ai, and J. Popelek (2000): "Board-level 2-D data-capable optical interconnection circuits using polymer fiber-image guides," Proceedings of the IEEE, vol. 88, no. 6, pp. 794-805.

Koonen, T. (2006): "Fiber to the home/fiber to the premises: what, where, and when?" Proceedings of the IEEE, vol. 95, no. 5, pp. 911-934.

Mazor, S. (1995): "The history of the microcomputer—invention and evolution," Proceedings of the IEEE, vol. 83, no. 12, pp. 1601-1608.

Roberts, B. P., and C. Sandberg (2011): "The role of energy storage in development of smart grids," Proceedings of the IEEE, vol. 99, no. 6, pp. 1139-1144.

Ross, P. E. (2011): "Top 11 technologies of the decades," IEEE Spectrum, vol. 48, no. 1, pp. 27-63.

Uman, M. A. (1988): "Natural and artificially-initiated lightning and lightning test standards," Proceedings of the IEEE, vol. 76, no. 12, pp. 1548-1565.

Zahn, M. (1979): Electromagnetic Field Theory: a problem solving approach, John Wiley & Sons, Inc.

Zakraoui, J., M. Saleh, S. Al-Maadeed, and J. M. Alja'am (2021): "Arabic machine translation: A survey with challenges and future directions," IEEE Access, vol. 9, pp. 161445-161468.

Zavadil, R., N. Miller, A. Ellis, E. Muljadi, P. Pourbeik, S. Saylors, R. Nelson, G. Irwin, M. S. Sahni, and D. Muthumuni (2011): "Models for change," IEEE Power & Energy Magazine, vol. 9, no. 6, pp. 86-96.

索引

A

AA battery *235*
AAA *237*
AAA battery *235*
abbreviation *2*
absolute participial construction *51*
abstract *3, 6, 9, 153, 214*
AC *46, 47, 235*
academic journal *2*
accept *138, 145, 213, 214, 226, 228*
accusative *207*
acknowledgment *3, 23*
acronym *2, 78*
active power *147*
active voice *29*
addition *66, 98*
address register *181, 183*
adjustable-speed system drive *166 167*
advanced metering infrastructure *160, 163*
advanced registration *215*
affix *207*
agree with *20, 56, 89, 110*
AI *203*
aim *12, 80, 160*
air carriage *224*
allographic *207*
alloy *57, 144*
alphabetical order *26*
alternating-current *46, 146*
American Automobile Association *237*
AMI *160, 162, 163*

ammeter *47, 156, 157*
ampere *41, 184*
Ampere's law *44, 134, 136, 137, 140*
amplification *2, 78, 158*
analog to digital converter *159, 160*
annual *170, 173*
anonymous *23*
antenna *152, 154, 189, 190, 191, 192*
antimony *176, 177*
apparatus *145, 146*
apparent power *145, 146*
appendix *3, 23*
application *55, 159, 160, 171, 172, 186, 187, 193, 195, 227, 230*
applied voltage *82, 92, 143, 147, 178*
apply *54, 55, 63, 82, 92, 107, 143, 147, 150, 151, 158, 178, 184, 209, 231*
arithmetic sequence *102*
arsenic *176, 177*
Artificial Intelligence *201, 202, 203, 207*
as far as *48, 193*
associate editor *213*
asynchronous transfer mode *195*
ATM *195, 236*
attenuate *7, 10, 79*
attenuation *7, 8, 10, 16, 18, 19, 22, 25, 26, 31, 32, 34, 46, 77, 82, 84, 85, 87, 90, 91, 107, 120, 121, 154*

author *3, 5, 6, 14, 27, 29, 62, 74, 75, 113, 115, 116, 212, 213, 219, 221, 225, 226*
automated teller machine *236*
autonomously *164, 165*
auxiliary verb *34*
avalanche effect *178*
AVOmeter *157, 158, 159*
award *27, 187, 230*
axial *7, 8, 18, 19, 22, 87, 89, 91, 121*

B

Bachelor of Science degree *27, 231*
background *74, 77, 79*
back issue *223*
back-to-back *180, 181*
bar graph *92*
based on *60, 107, 159, 187, 203, 209, 227*
battery *157, 173, 174, 178, 235*
battery runtime *174*
BCD *181, 183*
be able to *35, 36, 62, 76, 137, 171, 203*
be aware of *38*
be capable of *36, 58, 170, 174, 203*
be concerned with *129, 130*
be defined as *128, 129, 130, 132, 145*
be dependent on *57, 133*
be directly proportional to *104, 143*
be going to *35, 113, 197, 221*
be independent of *57, 132, 133,*

242

138, 145

be inversely proportional to *63, 104, 143*

be proportional to *104, 132, 143, 171*

be related to *134, 136, 206*

be subjected to *144*

because *7, 22, 47, 80, 87, 132, 137, 169, 172, 176, 181, 185, 190, 191, 197*

because of *54, 59, 60, 137, 169, 170, 171, 172, 178, 191*

BERT *209, 210*

biennial conference *214*

binary-coded decimal *182*

biography *3, 27*

bipolar junction transistor *53, 180, 181*

blackout *164, 165*

boron *176, 177*

brace *99*

breakdown voltage *178, 179*

broadband *194, 195*

broken line *93*

B.S. degree *27, 231*

B.Sc. degree *27, 231*

built-in *43, 181, 183*

bullet *67*

by definition *134, 136*

by means of *61, 196*

by use of *61*

C

cable television *196*

calculate *16, 45, 46, 50, 94, 107, 134, 146, 148*

calculation *19, 40, 50, 84*

calibrate *155*

call for papers *214*

can *8, 36, 37, 39, 46, 47, 50, 52, 60, 63, 66, 69, 70, 76, 80, 81, 82, 83, 84, 85, 87, 89, 92, 93, 94, 111, 120, 121, 129, 133, 134, 140, 152, 154, 157, 159, 163, 166, 168, 171, 172, 178, 180, 190, 191, 192, 196, 197, 200, 201, 202, 203, 205, 208*

capable *36, 57, 170, 174, 203*

capacitance *18, 20, 89, 144, 145, 147, 152*

capacitor *55, 63, 145, 146, 147, 150, 151, 154*

capital *4, 101*

caption *15*

carbon-polyethylene compound *10*

cascaded *154*

case marking *206, 207*

cathode ray oscilloscope *158, 159*

CATV *195, 196*

cf. *25*

chairperson *75, 76, 112*

characteristic impedance *49, 94, 154, 155*

charge *20, 48, 49, 65, 89, 132, 133, 135, 137, 138, 154, 169, 171, 176, 178*

charge carrier *176, 177, 178*

charge conservation *137, 140*

chatbot *204, 205, 209, 210*

checking account *236*

chronological order *26*

circuit breaker *161, 163*

circuital integral *101*

circular *13, 191*

circular polarization *191, 192*

cloud discharge *170*

cloud-to-air discharge *169, 170*

cloud-to-cloud discharge *170*

cloud-to-ground lightning *170*

coal-fired power plant *52, 171, 173, 187*

coaxial *46, 51, 89, 152, 153*

code division multiplexing *195*

colloquium *214*

colon *63, 65*

column *94, 106*

combination *102, 103*

common log *98*

compare *25, 43, 55, 56, 62, 108, 159, 187*

compared to *55, 187*

compared with *56*

comparison *56, 103*

complete *37, 117, 137, 138, 140*

complex number *104*

component *48, 130, 131, 183, 193*

compound figure *18*

compressed air *171, 173*

computation *7, 8, 13, 14, 24, 41, 42, 43, 47, 55, 108, 109, 225, 226, 227, 228, 230*

compute *7, 8, 11, 20, 29, 30, 31, 43, 55, 56, 57, 81, 82, 85, 87, 89, 92, 93, 108, 109, 120, 121, 228*

computer code *189, 190*

concatenative *206, 207*

conclusion *3, 22, 33, 40, 74, 90, 107, 115, 121*

conductance *19, 22, 46, 86, 154*

conduction *19, 86, 91, 176*

conduction current *19, 86, 91*

conductivity *7, 8, 10, 13, 15, 16, 17, 18, 19, 22, 31, 40, 57, 64, 77, 80, 82, 83, 84, 85, 87, 88, 89, 90, 91, 92, 93, 107, 120, 121, 123*

conductor *10, 13, 14, 15, 16, 19, 20, 30, 31, 45, 51, 59, 61, 62, 65, 66, 70, 79, 80, 82, 89, 94, 120, 141, 144, 152, 153, 175*

conference *112, 214, 225, 226*

conjunction *45, 53, 155, 156*

connect *47, 55, 63, 146, 147, 148, 150, 151, 152, 154, 156,*

178, 180, 181, 201, 203
connecting lead   180, 181
connection   151, 159, 180, 181, 197
conservation of charge   137, 138
consonant   206, 207
consumed power   147
contour   13
contrary to   62
contribution   48, 130, 131
convention   214
conventional   58, 163, 164, 181
coordinate conjunction   45
copyright   220
copyrighted figure   220
core network   195
corona discharge   41, 43, 44
corpus   208, 209
corresponding   16, 20, 43, 55, 57, 89, 108, 206, 223, 228
cost-effectiveness   193, 194
CO_2 emission   160, 163
could   34, 35, 36, 37, 107, 108, 109, 110, 114, 118, 138, 171, 219, 222, 225, 226, 228
covalent bond   176, 177, 178
crosslinked polyethylene insulated cable   10
cross product   129
cross-section   13, 52, 80, 120
cube root   97
cumulonimbus   169, 170
curl   50, 105, 133, 134, 136, 137, 140
current density   134, 137
curriculum vitae   25, 229, 231
curtailment   163, 165
C.V.   25, 228, 231
cybersecurity   201, 202

D

dart leader   5
dash   63, 64, 65, 93, 98
dashed-dotted line   93
dashed line   93
data communication   7, 10, 70, 79, 193
DC   46, 47
debit card   236
decimal   96, 183
deep learning   201, 202, 203, 208, 209
deficiency   176
deficient   176, 177
definite article   42
definite integral   101
deflection   156, 157, 158
del   105, 129
del operator   129, 130
denominator   97
Department of Motor Vehicles   237
depend   82, 107, 121, 175, 176, 206
dependence   16, 17, 31, 77, 82, 83, 84, 88
dependent   57, 133
depletion layer   49, 178, 179
deposit   236
depth   7, 19, 22, 87, 89, 121
derivation   206, 207
derivative   100, 101, 129, 130, 141
derive   54, 60, 134, 138
desktop calculator   181, 182
determinant   106
determine   19, 88, 134, 145, 147, 151, 154
diacritic   206, 207
diamagnetic   149
dielectric   66, 70, 145, 190, 191, 192
dielectric constant   190
differential   100, 129, 130
digital instrumentation   159, 160
digital multimeter   159, 160
digital oscilloscope   159, 160
digital subscriber line   199
digital subscriber loop   197, 199
digital transformation   207, 209
diode   55, 62, 175, 178, 184, 186
diphthong   206, 207
direct-current   46, 53, 143, 150, 151, 155, 157, 165, 166
directivity   191, 193
discharge   32, 33, 41, 43, 44, 51, 52, 78, 169, 170, 171, 172
discrepancy   58
discuss   18, 19, 32, 77, 85, 87, 110, 121, 195, 209
discussion   3, 18, 74, 85, 107, 113, 115, 117, 121
displacement current   19, 86, 137, 138, 140
distance   15, 16, 30, 31, 42, 81, 82, 83, 85, 87, 92, 93, 120, 129, 193, 196
distributed generation   160
diurnal   171, 172, 173
divergence   105, 134, 135, 136, 137, 138, 141
division   99, 183, 195, 196
Doctor of Philosophy   27, 232
Doctor of Science   232
donate   176, 177
doping   176, 177
dot product   128
dotted line   93
double A battery   235
double integral   101
double space   2
drift   49, 178, 179

driver license office *237*
driver's guide *237*
DS-160 *234*
DS-2019 *233, 234, 237*
DSc *232*
DSL *197, 198, 199*
due to *59, 60, 66, 86, 87, 121, 132, 138, 164, 177, 178, 196, 205*

E

ecology *168, 169*
eddy current *47, 65, 148, 149*
editor-in-chief *213*
efficiency *63, 184, 186, 187, 190, 193*
e.g. *25, 163, 166, 169*
electric discharge *51, 52, 78, 169*
electric field *20, 48, 50, 65, 66, 89, 132, 133, 137, 140*
electrical breakdown *10, 79*
electrical circuit *143, 149*
electrical discharge *78, 170*
electricity *145, 160, 161*
electromagnetic energy *152, 153*
electromagnetic field *47, 52, 128, 137, 225, 227, 230*
electromagnetic wave *19, 137, 189*
electromagnetics *112, 128, 189, 227, 229, 231*
electromotive force *62, 148, 149, 150*
electron *58, 158, 176*
electron beam *158, 159*
electronics *152, 175, 193, 216, 220, 222, 223, 224, 225, 227, 229*
electronics device *175*
elliptical *191, 192*

e-mail *116, 197, 220, 222, 223, 224, 226, 227, 229, 231*
en masse *197, 199*
end user *163, 194, 195*
energy *20, 34, 49, 51, 89, 141, 143, 146, 148, 150, 152, 153, 160, 161, 163, 164, 165, 167, 170, 171, 172, 174, 178, 187*
energy storage unit *171, 173*
equipment *145, 146, 163, 166, 167, 168, 170*
equivalent *10, 34, 230*
error *36, 38, 69, 158, 205, 222*
et al. *10, 24, 25, 26, 32, 33, 68, 164, 166, 168, 174, 193, 202, 203, 205, 206, 208, 209*
etc. *25, 181, 206, 209*
Ethernet *195, 196*
Euler's formula *44, 104, 105*
evaluate *16, 59*
except *62, 185*
exception to the rule of sequence of tenses *33*
expand *140, 142*
expression *54, 60, 89, 95, 128, 132, 137, 143, 150, 155, 160, 167, 170, 175, 181, 184, 189, 193, 197, 201, 204, 207*
extended hypertext markup language *197, 199*
extensible markup language *197, 199*

F

fabrication *191, 192*
false alarm rate *202, 203*
FDTD *4, 7, 8, 9, 11, 12, 13, 14, 20, 22, 23, 29, 30, 31, 32, 33, 35, 41, 42, 43, 44, 55, 57, 77, 78, 79, 80, 81, 83, 85, 86, 87, 89, 90, 91, 92, 93, 110, 120, 121, 122*
feasibility *165, 166*

feed line *191, 192*
ferromagnetic *47, 148, 149*
fiber to the home *195*
fiber to the premise *195*
finite-difference time-domain method *3, 4, 7, 8, 9, 11, 25, 26, 29, 41, 42, 43, 44, 55, 77, 78, 120*
fluorescent *184, 185, 187*
FM *153*
for the purpose of *60, 61*
for the sake of *60*
force *49, 62, 132, 148, 149, 150, 171, 181, 200, 201*
forerunner *184, 186*
forward biased *178, 179, 180*
fraction *96, 97, 171, 190, 192*
fractional part *96*
free space *140, 142*
frequency *7, 8, 10, 11, 14, 16, 19, 22, 34, 63, 79, 82, 87, 92, 121, 147, 148, 152, 153, 158, 163, 164, 165, 195*
frequency modulation *152, 153*
FTTH *195*
FTTP *195*
full wave rectified current *155, 156*
full paper *213, 215*
future tense *32*

G

galvanometer *157*
GAN *209, 210*
gas turbine *171, 173*
gaussian mixture model *209, 210*
gender neutral *76*
generative adversarial network *209, 210*
generative AI *208, 209*
genitive *206, 207*

geometric sequence *102*
germanium *175, 176*
gerund *55*
give credit *25*
GMM *209, 210*
goal *12, 52, 80, 198, 201*
governor control *164, 165*
Google Translate *69, 70, 205*
gradient *129, 130, 134, 136*
grant *24, 196*
graph *92, 93, 116*
Greek letter *99, 223*
group preposition *54, 59*
guide for authors *2, 62, 215*

H

habilitation *232*
had better *40*
handout *123*
handset *184, 186*
Harvard referencing system *25*
HDTV *186, 188, 189*
Helmholtz's theorem *134, 135, 136*
HEMS *160, 162, 163*
hertz *79, 82, 83, 84, 85, 86, 87, 89, 90, 92, 108, 120, 121*
hidden layer *203, 204*
hidden Markov model *209, 210*
high current *47, 170*
high definition television *186, 189*
high-pass filter *155*
high-resource language *204, 205*
high voltage direct current *165, 166*
h-index *218*
hole *176, 177, 178*
HMM *209, 210*
home energy management system *160, 163*

homogeneous *140, 142*
hook *197, 199*
horizontal axis *82, 84, 92, 93, 116, 120*
HVdc *165, 166*
hyphen *63, 64*
hypothesis *137, 140*

I

I-901 SEVIS *234*
ibid. *25*
ICT *161, 162, 163*
identical *57, 104, 148*
i.e. *25, 144, 149, 155, 158, 206*
IEC *235*
IEEE *184, 216, 217*
IEEJ *184, 217, 238*
IET *216*
inappropriate authorship *5*
immediately *20, 89, 150*
impact factor *216*
impedance *42, 45, 49, 94, 146, 149, 150, 152, 154, 155*
impurity *176, 177*
in addition to *62, 148, 149*
in comparison with *56*
in conjunction with *53, 155, 156*
in contrast to *62, 203*
in excess of *175, 176*
in inverse proportion to *63, 104*
in parallel with *47, 55, 63, 146, 147, 156*
in parentheses *129, 130*
in place of *61*
in press *24, 231*
in response to *164, 165*
in series with *55, 63, 147, 150, 151, 156*
in spite of *63*
in the vicinity of *31, 132, 133*
incandescent *184, 186, 187*
incident *154, 155*

indefinite arcticle *41*
indefinite integral *101*
independent *57, 132, 133, 138, 145*
index term *3, 9*
indium *176, 177*
inductance *20, 45, 46, 50, 89, 147, 148, 151, 152*
inductor *150, 151*
industrialized society *160, 162*
inequality *103*
infinitive *61*
infinity *102*
information and communication technology *161, 163*
inflection *206, 207*
inject *15, 16*
inner product *105, 128*
install *80, 120, 145, 174*
instead of *54, 61*
instructions to authors *2, 212*
instrument *46, 47, 53, 61, 155, 156, 157, 169*
insulating layer *10, 13, 19, 20, 70, 79, 80, 88, 89, 120*
insulator *19, 45, 59, 66, 88, 144, 175, 176*
integral *50, 58, 59, 100, 101, 105, 133, 141*
interaction *161, 189, 190*
international conference *112, 214*
Internet *197, 199, 200, 201, 208*
Internet Protocol *199, 201*
intracloud discharge *169, 170*
intrinsic semiconductor *177*
introduction *3, 9, 30, 160, 209*
intrusion detection system *201, 202*
invitation letter *226*
IP *199, 200, 201*
iron core *62, 148*

J

Japan Society for the Promotion of Science *27, 230*
jargon *2, 78*
justification *197, 199*

K

Kelly blue book *237*
kelvin *42, 96*
keyword *3, 9*
kinetic energy *164, 165*

L

laboratory *28, 174, 189, 227, 228*
LAN *124, 125*
Laplacian *101, 135, 136*
large scale integrated chip *181, 182*
laser *2, 76, 78, 193, 194*
leading *19, 147, 148, 193, 197, 199*
LED *184, 185, 186, 187, 188*
leftmost *94*
legacy system *161, 163*
legend *15, 18*
letter *206, 213*
lexical *206, 207*
ligature *206, 207*
light *2, 14, 35, 37, 62, 78, 112, 113, 115, 137, 140, 169, 170, 184, 185, 186, 187*
light-emitting diode *62, 184, 186*
lighting *62, 185, 187*
lightning *5, 31, 32, 33, 35, 40, 48, 52, 55, 59, 61, 112, 115, 169, 170, 219, 221, 223, 225, 226, 227, 228, 230, 231*
lightning flash *5*
Li-ion (lithium-ion) battery *174, 175*
limit *14, 102, 108, 113, 145, 161, 171, 181, 193, 203*
line graph *92*
line integral *50, 133*
linear polarization *191, 192*
liquid crystal display *186, 189*
local area network *55, 194*
logarithm *97*
logic chip *181, 182*
longevity *185, 186*
LSI chip *181, 182*
lumen *184, 186, 187*

M

Machine Learning *201, 202, 203, 207*
Machine translation *204, 205, 206*
magnetic field *20, 31, 48, 89, 132, 134, 137, 140*
magnitude *16, 19, 31, 49, 82, 83, 84, 86, 88, 92, 120, 128, 130, 132, 146, 154, 156, 166, 170, 178*
main clause *8*
majority carrier *49, 178, 179*
malicious activity *201, 202*
manufacturer *155, 157, 170*
manuscript *213, 214, 219, 222*
marked *7, 16, 22, 82, 158*
Master of Science degree *27, 231*
mathematical expression *95*
mathematical induction *102*
matrix *106*
Maxwell's equations *44, 60, 137, 138, 140, 228*
may *34, 38, 48, 51, 53, 67, 76, 134, 149, 152, 155, 157, 158, 163, 164, 167, 168, 169, 170, 191, 195, 200, 203, 206, 222*
mechanism *5, 11, 32*
meeting *214*
memory *14, 75, 108, 159, 181, 182, 183*
memory chip *181, 182*
Meta *208, 209*
method *3, 7, 8, 9, 11, 12, 13, 22, 25, 26, 29, 30, 31, 32, 33, 34, 35, 41, 43, 44, 47, 52, 54, 55, 56, 57, 62, 68, 74, 77, 80, 90, 92, 110, 120, 170, 189, 190, 209, 226, 227, 228, 230, 231*
method of moments *56, 68, 189, 190, 230*
microstrip antenna *190, 191, 192*
middle *94, 206*
might *7, 10, 34, 38, 79*
milliammeter *156, 157*
mimic *163, 203, 204*
minimum balance *236*
Ministry of Education, Culture, Sports, Science and Technology (MEXT) of Japan *24*
minority carrier *178*
mitigate *10, 60, 79*
ML *202, 207, 208*
mobile phone *184, 186*
monotonic *16, 31, 82, 84*
morphological *206, 207*
moving-coil instrument *53, 155, 156*
M.S. degree *27, 230, 231*
M.Sc. degree *27, 230, 231*
multilingual *204, 205*
multiplex *181, 183*
multiplexing *195*
multiplication *99, 183*
must *2, 37, 39, 40, 49, 132, 134, 135, 137, 138, 156, 163, 164, 167, 168, 199, 203*

247

N

NaS battery *174*
natural language processing *204, 205*
natural log *97*
natural science *232*
neural network *203, 204, 208*
network administrator *200, 201*
network topology *195*
N-gram language model *206, 210*
nimbostratus *169, 170*
nominal impedance *155*
nominative *206, 207*
norm *105*
normal component *131*
no-show *116*
notebook-size personal computer *43*
noughties *184, 186*
NN *204*
nth root *97*
nuclear power plant *53, 171, 173*
numerator *96*

O

objective *11, 74, 77, 79, 80, 120, 168, 229*
offshore *165, 166*
ohm *42, 94*
Ohm's law *143*
on account of *59*
on behalf of *61*
On-demand Oral Presentation *124, 125*
Online Oral Presentation *124*
on the basis of *14, 60, 195*
ONU *196*
open-circuited *150*
optical fiber *193, 194*
optical network unit *196*
optoelectronics device *184*
oral presentation *74, 214*
orthographical *206, 207*
outage *170, 173*
outer product *105, 129*
outline *74, 77, 118*
overhead transmission *144, 168, 169*
owing to *10, 59, 79*

P

paragraph *6*
parenthesis *99*
participial construction *45, 49, 51*
passive voice *27, 28*
past participle *50*
past tense *32*
patent *190, 192*
PCB *181, 182, 183*
penetration *7, 19, 22, 87, 89, 121, 161*
penetration depth *7, 19, 22, 87, 89, 121*
perform *24, 32, 33, 58, 79, 157, 159, 203, 228*
periodic *158, 170, 173*
permeability *19*
permit *24*
permittivity *13, 18, 19, 80, 88, 89, 120*
permutation *102, 103*
perpendicular to *128, 129, 130*
personal check *236*
personal computer *14, 37, 43, 183*
phase angle *146, 147*
Ph.D. *27, 39, 75, 120, 220, 221, 223, 224, 225, 227, 228, 229, 230, 232*
Ph.D. dissertation *232*
PHEV *160, 162, 163*
Philosophiae Doctor *27, 232*
Phonemic *206, 207*
Phonolongical *206, 207*
phosphorus *176, 177*
pie chart *92*
Plagiarism *68, 71*
Planck's constant *44*
PLC *4, 7, 8, 9, 10, 11, 12, 22, 23, 29, 33, 70, 77, 78, 79, 80, 90, 91, 121, 122*
plug-in-hybrid electric vehicle *160, 163*
point charge *49*
point-to-multipoint *195*
point-to-point *195*
poster presentation *118*
power cable *3, 4, 5, 7, 9, 10, 11, 13, 15, 16, 18, 19, 22, 25, 26, 29, 31, 33, 52, 70, 74, 75, 76, 77, 79, 80, 82, 85, 87, 90, 92, 93, 109, 120, 123*
power distribution line *10, 152, 153*
power distribution network *160, 162*
powered by *209*
power factor *145, 146, 147*
power line communication *3, 4, 5, 9, 10, 25, 64, 70, 75, 77, 78, 120*
power plant *53, 165, 166, 171, 173, 187*
power station *160, 163*
power system *22, 91, 121, 160, 163, 165*
poynting vector *141, 143*
premultiply *129, 130*
preposition *54, 59*
present participle *50*
present perfect tense *33*
present tense *30*

prime mover *164, 165*
printed circuit board *181, 183*
private communication *24*
probability *102, 103*
proceedings *2, 25, 26, 218*
process-control equipment *166, 167*
processor *14, 108, 183*
program *14, 59, 111, 189*
proof *222*
proofreading *69, 214*
propagation characteristic *7, 8, 11, 22, 29, 33, 77, 79, 90, 120*
property *145, 220*
provided *143, 174*
publication *39, 212, 221, 224, 228, 231*
publication fee *39*
pulse *14, 15, 16, 17, 46, 55, 81, 82, 92, 93, 107*
pumped hydroelectric *171, 172, 173*
pumped storage hydropower *174, 175*
pure semiconductor *176*
purity *175, 176*
purpose *12, 60, 61, 69, 80, 178*

Q, R

Q-factor *147*
radial *7, 8, 18, 19, 22, 86, 87, 91, 121*
radio frequency *152, 153*
radio transmitter *152, 153*
radius *13, 18, 31, 80, 94*
RAM *108, 181, 182*
random-access memory *108, 182*
range *7, 8, 10, 13, 19, 79, 80, 88, 94, 120, 157, 159, 166, 190, 193, 213*
rating *145, 146*

reactive power *146*
read-only memory *182*
real-time information *161, 163*
recipient *27*
reciprocal *97*
rectangular *190, 191*
rectifier *53, 155, 165, 157*
recurring decimal *96*
reference *3, 24, 144, 159, 160, 178, 225, 231*
referred to as *161, 163, 190, 191, 192*
reflect *154, 155, 168*
refrigerator *172, 174*
regardless of *63*
reinforcement learning *203, 204, 208*
relation *10, 28, 184, 186, 187*
relative permittivity *13, 18, 19, 80, 89, 120*
relative pronoun *45, 51*
reliability *163, 165*
repelling *178, 179*
research fellow *27, 230*
resistance *18, 45, 46, 47, 59, 60, 143, 144, 147, 148, 151, 152, 154, 156, 157, 159*
resistivity *175, 176*
resistor *28, 55, 63, 64, 144, 146, 150, 151, 152, 156*
resonant frequency *147*
respectively *16, 19, 27, 88, 197*
resubmit *213, 214*
result *3, 10, 15, 19, 30, 31, 55, 56, 69, 74, 77, 81, 82, 84, 107, 108, 109, 120, 121, 128, 130, 137, 163, 165, 176, 177, 178, 197, 201, 202, 208, 228*
résumé *25, 231*
reverse biased *178, 179, 180*
reviewer *23, 36, 69, 213*
revise *213, 214*

revision *54, 213*
right-hand side *137, 138, 140*
rightmost *94*
RMS *156*
ROM *181, 182*
root-mean-square *155, 156*
rotation *105*
rotational energy *164, 165*
rotationally symmetric *13*
row *94, 106*
rule of sequence of tenses *33*

S

sampling rate *159, 160*
satellite link *197, 199*
sawtooth generator *158, 159*
SCADA *161, 162, 163*
scalability *161, 163*
scanning capability *191, 192*
scanning electron microscope *58*
scientific electronic calculator *50*
scientific journal *2, 212*
selector switch *157*
SEM *61*
semantic *206, 207*
semicolon *63, 67*
semiconducting layer *3, 4, 5, 7, 8, 9, 10, 11, 13, 18, 19, 20, 22, 25, 26, 29, 70, 75, 77, 79, 80, 82, 85, 86, 87, 88, 89, 91, 92, 107, 120, 121*
semiconductor bandgap voltage reference *159, 160*
semi-supervised learning *203, 204, 208*
sequence *33, 102, 159, 200, 209*
serial arithmetic *181, 182*
series *46, 55, 63, 102, 147, 150, 151, 154, 156, 157, 159*
shall *34, 35, 39, 40*

short-circuited *150*
short paper *213*
should *34, 39, 40, 62, 137, 152, 175, 197*
shunt *156, 157, 167, 169*
shunt reactor *167, 169*
significant *7, 8, 10, 16, 18, 19, 22, 34, 82, 85, 87, 90, 91, 96, 120, 121, 164, 208, 209*
silicon *58, 61, 175, 176, 178*
similar *56, 87, 135, 152, 158, 159, 186, 208*
since *7, 8, 10, 19, 20, 22, 45, 47, 50, 79, 89, 104, 121, 134, 138, 141, 145, 148, 156, 163, 165, 167, 174, 178, 184, 186, 191, 198, 209*
sinusoidal *155, 156*
skin effect *60*
slack node *166*
slide *74, 77, 81, 84*
smart grid *161, 163*
SMES *171, 172, 173*
SMIL *197, 198, 199*
social security number *237*
sodium sulfer battery *174*
solid line *93*
source code *208, 209*
spatial derivative *129, 130*
speed *14, 18, 137, 140, 158, 166, 167, 178, 197*
speed of light *14, 137, 140*
spinning reserve *172, 173*
square *14, 88, 93, 97, 98, 99, 100, 101, 103, 158, 184, 191*
square bracket *99*
square root *88, 97, 98, 103*
SSN *237*
stability *14, 172*
stabilization *178, 179*
staircase-approximated *13*
steady-state *151*

Stokes' theorem *50, 133*
submission *68, 69, 196, 212*
subordinate clause *8*
subordinate conjunction *45*
subscript *98*
subset *201, 202*
substation *161, 163, 174*
substrate *190, 191, 192*
subtraction *66, 98*
summary *3, 22, 33, 74, 90*
supercomputer *24*
superconducting magnet *171, 173*
superconducting magnetic energy storage *171, 173*
superscript *98*
supervised learning *203, 204, 208*
supervisory control and data acquisition *161, 163*
surface *10, 13, 19, 20, 48, 50, 57, 70, 79, 89, 105, 130, 131, 133, 141, 171, 189, 173, 194*
susceptance *19, 22, 86*
sustainable *160, 163*
symmetry *138, 140*
symposium *214*
synchronization *196*
synchronized multimedia integration language *197, 199*
syntactic *206, 207*
syntactical *206, 207*

T

table *44, 92, 94*
tangent *131*
tangential component *130, 131*
task force *200, 201*
TDMA *196*
telephone line *152, 153*
temperature *36, 42, 45, 57, 58, 59, 143, 144, 145, 161, 175, 176, 187*
tetrahedral *176, 177*
textbook *50*
therefore *19, 20, 22, 88, 89, 104, 107, 108, 121, 161, 204*
thermal expansion *10, 70, 79*
thermometer *36, 58*
thundercloud *5, 32, 33, 48, 169*
tilde *98*
time constant *7, 18, 20, 22, 86, 89, 121, 151, 152*
time division multiple access *196*
time increment *14*
time-interleave *196*
time varying *48, 137, 140*
title *2, 3, 74, 75, 113, 115, 219, 221, 224, 225, 226, 230*
toxic mercury *185, 186*
transformer *145, 148, 149, 174, 208, 209, 210*
transient *25, 26, 52, 150, 152, 169*
transient phenomenon *150*
transistor *53, 58, 61, 180, 181, 183*
transmission congestion *163, 165*
transmission line *46, 49, 51, 144, 145, 152, 153, 154*
transmission system operator *165, 166*
transparent *158, 159*
trapezoidal rule *59*
traveling wave *123, 154*
trigonometric function *104*
triple A battery *235*
TSO *165, 166*
turbine generator shaft *164, 165*
two-dimensional cylindrical

coordinate system *11*
typo *222*
typographical error *222*

U

UHV *168, 169*
ultrahigh-voltage *167, 169*
unit normal *130, 131*
unit normal vector *131*
unit tangent vector *131*
unsupervised learning *203, 204*
upper case *4, 101*
upstream packet *196*
USB plug-in memory stick *75*
useful *34, 110, 213*
user-oriented software *189, 190*
utility *172, 173, 174, 236*

V

VA *147, 174*
vacuum *18*
valency electron *176, 177*
valuable *23*
value *8, 13, 19, 31, 45, 46, 47, 50, 56, 80, 84, 87, 88, 89, 94, 120, 121, 132, 134, 147, 151, 155, 156, 159, 164, 166*
Vancouver referencing system *25*
VAR *147*
variable frequency *147*
variant *206, 207*
vector *47, 66, 105, 128, 129, 130, 131, 132, 134, 135, 136, 137, 140, 141, 149*
vector potential *134, 136*
vector product *128, 129*
versus *25*
vertical axis *82, 84, 92, 93, 116, 120*
viewgraph *75*
visa *226*
visual pollution *167, 168, 169*
voltage sag *166, 167*
voltage source converter *165, 166*
voltage stabilization *178, 179*
voltampere *145, 147*
voltampere reactive *147*
voltmeter *156, 157*
volume integral *133*
vowel *206, 207*
vs. *25*
VSC *165, 166*

W

wall outlet *235*
watt *145, 147, 184, 187*
waveform *15, 16, 30, 43, 57, 81, 85, 87, 92, 93, 120, 158*
waveguide *51, 152, 153*
wavelength *187, 190, 192, 195*
web page *52, 197, 198*
well-nigh *198, 199*
while *16, 17, 82, 131, 144, 160, 163, 200*
whole number part *96*
will *19, 32, 34, 35, 49, 57, 75, 76, 77, 91, 113, 114, 115, 117, 121, 123, 134, 137, 160, 161, 164, 171, 178, 184, 185, 187, 193, 194, 197, 198, 200, 203, 209, 221, 228*
wind farm *174*
wind power plant *165, 166*
wireless communication *189*
wire-type linear antenna *189, 190*
with the help of *61*
withdraw *236*
World Wide Web Consortium *197, 199*
would *20, 23, 34, 35, 36, 76, 77, 79, 88, 90, 110, 111, 115, 117, 120, 121, 132, 137, 138, 140, 144, 150, 163, 167, 181, 183, 187, 198, 213, 220, 221, 222, 224, 225, 228,*
would like to *23, 24, 35, 76, 77, 79, 88, 90, 115, 117, 121, 224, 225, 228*

X, Z

XHTML *197, 198, 199*
XLPE cable *10, 11*
XML *197, 199*
X-ray *5*
Zener effect *178, 179*

251

索引

あ

アップストリームパケット 196
アドレスレジスタ 182, 183
アナログ-デジタル変換器 159, 160
アメリカ電気電子学会 216, 220, 228
アルファベット順 26
アンチモン 177
アンテナ 153, 154, 189, 190, 191, 192
アンペア 36, 44, 135, 136, 137, 139, 142, 146, 157, 174, 185, 188
アンペアの法則 44, 135, 136, 137, 139, 142
圧縮空気 172, 173
安定化 165, 179
安定性 23

い

イーサネット 195, 196
イギリス工学技術学会 216
インジウム 177
インターネット 197, 198, 200, 201, 202, 209, 234, 236
インターネットプロトコル 200, 201
インダクタ 151, 152
インダクタンス 21, 46, 90, 143, 147, 148, 152
インパクトファクタ 212, 216, 217, 218
インピーダンス 42, 45, 49, 50, 94, 143, 147, 149, 150, 152, 154, 155
位相角 147
依存 17, 31, 57, 69, 78, 83, 85, 89, 107, 122, 133, 140, 177, 204, 207
著しい 7, 8, 9, 11, 20, 21, 23, 34, 80, 83, 86, 87, 91, 122, 205, 215
一致 8, 22, 33, 34, 38, 51, 56, 57, 58, 89, 90, 109, 140, 158, 188
一点鎖線 93
一方で 6, 17, 70, 169, 194
異綴語 207
意味論的な 207
異文字の 207
印加電圧 83, 93, 143, 179
陰極線オシロスコープ 155, 158, 159
印刷中 24, 231

う

ウェブページ 2, 52, 198, 199, 212, 213, 214, 215, 222, 233, 234, 237, 238
上付き文字 25, 98
渦電流 47, 65, 143, 149
右辺 98, 139, 140, 142
雲間放電 170
運動エネルギー 165
雲内放電 170

え

エックス線 5
エネルギー 21, 34, 49, 51, 90, 142, 143, 146, 149, 150, 153, 160, 162, 163, 165, 168, 170, 172, 173, 175, 179, 188
エネルギー貯蔵 170, 172, 173, 175
エンドユーザ 195
衛星中継 198, 199
液晶ディスプレイ 188, 189
N-gram 言語モデル 209, 210
円グラフ 92
円形 14, 192
円偏波 192
h-指数 218
n 乗根 97

F-1 ビザ 234
F-2 ビザ 234

お

オイラーの式 44, 104
オーム 142, 143
多くの言語を使える 205
大文字 4, 5, 41, 44, 66, 100, 101
温度 36, 42, 45, 57, 58, 59, 96, 143, 144, 146, 162, 175, 176, 188
温度計 36, 58
音韻論的な音素 207

か

ガスタービン 172, 173
会議参加事前登録 215
階段近似 14
回転 14, 50, 133, 135, 136, 139, 142, 165
回転エネルギー 165
回転対称 14
開放 150
概要 74, 77, 78, 118, 122
架橋ポリエチレン絶縁ケーブル 11
架空送電 145, 169
架空配電 145
各家庭までの光ファイバ化 195
各敷地までの光ファイバ化 195
学士号 27, 231
学術雑誌 2, 4, 7, 212, 216
拡張したハイパーテキストマークアップ言語 198, 199
拡張性 162, 163
格標示 207
確率 102, 210
隠れ層 204
隠れマルコフモデル 209, 210
掛け算 99
過去時制 30, 32, 33
過去分詞形 50, 149, 186

仮説　139, 140
価値　216, 227
括弧　26, 64, 99, 130
価電子　177
可動コイル形計器　53, 155, 156, 157, 159
過渡現象　150
過渡状態　152
可変速度システム駆動装置　167
雷　5, 31, 32, 33, 35, 40, 48, 49, 52, 55, 60, 61, 113, 115, 167, 170, 226, 227, 228
関係　11, 18, 27, 28, 41, 42, 45, 51, 52, 53, 54, 63, 67, 130, 132, 133, 135, 136, 139, 146, 185, 188, 207, 222, 232
関係代名詞　45, 51, 52, 53
監視制御・データ収集　162, 163
関数電卓　50
完成　37, 212

き

キーワード　2, 3, 4, 9, 71
キャパシタ　55, 63, 145, 146, 147, 151, 154
キャパシタンス　20, 21, 90, 143, 145, 147, 148, 152
ギリシャ文字　99, 100, 223
起因　20, 60, 86, 88, 122, 132, 140, 154, 179
機械学習　201, 202, 203, 204, 208
機械翻訳　69, 70, 71, 201, 204, 205, 206
基幹回線網　195
機器　146, 157, 167, 169, 170, 172, 173, 187, 194, 196, 235
起電力　62, 149, 150
基板　182, 183, 191, 192, 194
規約拡張マークアップ言語　199
逆数　97
逆方向にバイアスされた　179, 181
級数　102
給電線　192
寄与　48, 131, 146
行送り　199
強化学習　203, 204, 208
教科書　50, 140
教師あり学習　203, 204, 208
強磁性体　47, 148, 149
教師なし学習　203, 204, 208, 210
教授資格　232
共振周波数　147
行端ぞろえ　199
共有結合　177, 179
行列　106
行列式　106
許可　24, 36, 37, 38, 220, 234
極限　102
距離　17, 31, 42, 130, 193, 194, 195, 196, 236
均質　142
Q値　147

く

グラフ　15, 18, 75, 92, 93, 94, 116
空間微分　130
空乏層　49, 179
屈折　207
組合せ　102
雲放電　170
群前置詞　46, 54, 59, 60, 61

け

ケーブルテレビ　196
ケルビン　173
ゲルマニウム　175, 176, 177
計器　46, 53, 61, 155, 156, 157, 159
軽減　60
蛍光灯　185, 186, 188, 189
掲載　2, 24, 39, 184, 212, 213, 214, 215, 216, 218, 220, 221, 222, 223, 224, 225, 228, 238
掲載料　39, 214
計算　8, 9, 11, 12, 13, 14, 15, 16, 17, 18, 21, 22, 23, 24, 29, 30, 31, 33, 40, 41, 42, 43, 44, 45, 46, 47, 50, 52, 55, 56, 57, 58, 59, 61, 71, 80, 81, 82, 83, 84, 85, 86, 87, 89, 90, 93, 94, 108, 109, 122, 135, 147, 148, 182, 189, 190, 216, 228
計算機コード　190
携帯電話　185, 186
形態論的な　207
欠陥　177
決定　24, 164, 184, 189, 213, 214, 216, 226
結論　22, 33, 40, 91, 107, 111, 116, 122
研究員　27, 227, 229, 230, 233, 237
研究室　24, 28, 227, 228, 233, 236, 238
研究補助金　24
原稿　23, 69, 70, 71, 214
言語モデル　209, 210
現在完了時制　30, 33
現在時制　30, 32, 34
現在分詞形　50
原子力発電所　53, 172, 173
減衰　7, 8, 9, 11, 17, 18, 20, 21, 23, 31, 33, 34, 46, 70, 78, 80, 83, 85, 86, 87, 88, 91, 93, 107, 122, 154
原動機　165
検流計　157

こ

コア　47, 148, 149
コストパフォーマンス　185, 186, 194, 195

索　　引

コーパス　*208, 209*
コロナ放電　*41, 43, 44*
コロン（：）　*18, 63, 65, 66, 67, 68*
コンダクタンス　*21, 23, 46, 86, 154*
語彙的な　*207*
高域通過フィルタ　*155*
工業化社会　*161, 163*
合　金　*57, 144*
航空郵送料　*224*
貢　献　*5, 6, 218*
考　察　*3, 18, 20, 21, 74, 78, 85, 86*
合　字　*207*
公称インピーダンス　*155*
校　正　*68, 69, 156, 214, 222*
高精細度テレビ　*188, 189*
校正刷り　*214, 222*
光　速　*14, 139, 140*
広帯域　*192, 193, 194*
口頭発表　*74, 95, 106, 112, 113, 238*
高度計量インフラストラクチャ　*162, 163*
勾　配　*128, 130, 135, 136*
降伏電圧　*179*
構文的な　*207*
高リソース言語　*205*
効　率　*11, 62, 63, 68, 70, 162, 169, 173, 185, 188, 192, 194, 202, 203, 213*
交　流　*46, 53, 62, 143, 146, 148, 149, 156, 157, 233, 235*
小切手　*236*
国際運転免許証　*234*
国際会議　*2, 4, 7, 112, 124, 125, 214, 215, 224, 225, 226, 238, 239*
国際電気標準会議　*235*
誤警報率　*202, 203*
誤　字　*214, 222*
小文字　*66, 100, 101*

混合ガウスモデル　*209, 210*

さ

サイバーセキュリティ　*202*
サセプタンス　*21, 23, 86*
サブセット　*202*
サンプリングレート　*159, 160*
最低預金額　*236*
査読者　*23, 24, 25, 36, 68, 69, 213, 215*
差　分　*3, 7, 9, 12, 29, 41, 42, 43, 44, 55, 77*
三角関数　*104*
参考文献　*2, 3, 6, 12, 22, 24, 25, 26, 71, 225*

し

シリコン　*58, 61, 175, 176, 177, 179*
子　音　*42, 43, 207*
したがって　*2, 6, 21, 23, 25, 30, 60, 89, 90, 98, 107, 109, 112, 122, 132, 133, 136, 143, 144, 146, 153, 162, 199, 200, 205, 207, 216*
磁　界　*21, 31, 47, 48, 52, 55, 68, 89, 113, 132, 135, 138, 139, 140, 142, 143, 149, 189, 191, 192, 228*
視覚公害　*168, 169*
時間インタリーブ　*196*
時間刻み幅　*12, 14*
時間変化　*48, 138, 139, 140*
時間領域有限差分法　*3, 7, 9, 12, 29, 41, 43, 44, 55, 77*
軸方向　*8, 9, 20, 21, 23, 88, 89, 91, 122*
指向性　*188, 192, 193*
私　信　*24*
時制の一致　*8, 33, 34, 38, 140*
時制の一致の例外　*8, 33, 140*
自然科学　*232*

自然言語処理　*205, 210*
持続可能　*162, 163*
下付き文字　*98*
実現可能性　*166*
実時間情報　*162, 163*
実　線　*14, 93*
時定数　*8, 20, 21, 23, 86, 90, 122, 150, 151, 152*
時分割多重アクセス　*196*
社会保障番号　*237*
謝　辞　*3, 5, 23, 24*
遮断器　*162, 163*
周回積分　*101*
周期的　*172, 173*
自由空間　*142, 191, 192*
修士号　*27, 231*
縦　続　*154*
従属節　*8, 33, 34, 48, 51, 140*
従属接続詞　*45, 46, 47*
充　電　*125, 154, 173*
周波数　*7, 8, 11, 12, 14, 17, 21, 22, 23, 29, 34, 51, 60, 63, 70, 80, 83, 85, 88, 91, 93, 122, 147, 148, 153, 155, 158, 164, 165, 195, 235*
周波数変調　*153*
従　来　*58, 164, 165, 182, 204*
受賞者　*27*
主　格　*51, 60, 207*
主　節　*8, 33, 48, 51, 140*
受動態　*7, 27, 28, 29, 50*
寿　命　*185, 186, 188*
循環小数　*96*
瞬時予備力　*173*
純　度　*176*
順方向にバイアスされた　*179, 181*
順　列　*102*
賞　*27, 188, 230*
小　数　*95, 96*
少数キャリヤ　*179*
小数部分　*96*

索　引

招待状　*215, 226*
消費電力　*147, 188*
情報通信技術　*162, 163*
緒　言　*3, 9, 10, 12, 30, 32, 69, 71*
所有格　*44, 51, 52, 105, 133, 207*
真　空　*20*
人工知能　*71, 201, 202, 203, 208, 210*
進行波　*123, 154*
真性半導体　*177*
深層学習　*201, 202, 203, 204, 208, 209, 210*
深層学習モデル　*204, 210*
浸　透　*195*
侵入検知システム　*202*
振　幅　*14, 17, 83, 85, 93, 122, 167, 196*
信頼性　*162, 164, 165*
J-1 ビザ　*233, 234, 237*
J-2 ビザ　*233, 234*

す

スイッチ　*157, 159*
スーパーコンピュータ　*24*
スカラ積　*128, 130*
ストークスの定理　*50, 132, 133*
スマートグリッド　*160, 162, 163*
スライド　*74, 75, 76, 77, 78, 79, 80, 81, 82, 83, 84, 85, 86, 87, 88, 90, 92, 94, 98, 112, 116, 124, 125, 215*
スラックノード　*166*
随時アクセスメモリ　*182*
数学的帰納法　*102*
数　式　*65, 66, 95, 140*
数　列　*102*

せ

セミコロン（；）　*18, 63, 67, 68*
正弦関数　*156*
正　孔　*177, 179*
製　作　*52, 157, 192*

正書法的な　*207*
整数部分　*96*
生成 AI　*71, 201, 207, 208, 209, 210*
製造業者　*173*
生　態　*169*
成　分　*48, 131*
正方形　*14, 192*
整流回路　*53, 156*
石炭火力発電所　*53, 172, 173, 188*
積　分　*50, 59, 100, 101, 133, 142*
積乱雲　*170*
絶縁層　*11, 14, 21, 70, 80, 81, 89, 90, 121*
絶縁体　*21, 45, 59, 62, 66, 89, 144, 176*
絶縁破壊　*11, 70, 80, 179*
絶縁破壊電圧　*179*
接　辞　*207*
接　線　*131*
接線方向成分　*131*
接　続　*5, 45, 46, 47, 49, 50, 51, 55, 63, 67, 112, 123, 125, 146, 147, 148, 149, 150, 151, 152, 153, 154, 156, 157, 159, 162, 166, 179, 180, 181, 182, 194, 198, 199, 202, 204, 207*
接続詞　*5, 45, 46, 47, 49, 50, 51, 67*
接続リード線　*180, 181*
背中合せ　*180, 181*
先駆者　*186*
線グラフ　*92*
線状アンテナ　*189, 190*
線積分　*50, 133*
前置詞　*5, 46, 52, 54, 55, 56, 57, 58, 59, 60, 61, 65, 98, 133*
全波整流　*156*
専門用語　*2, 78*

そ

ソースコード　*208, 209*
相互作用　*190*
走査形電子顕微鏡　*58*
走査能力　*192*
送電系統運用者　*166*
送電混雑　*164, 165*
増　幅　*159*
即　座　*21, 90, 222*
速　度　*17, 18, 21, 22, 49, 78, 83, 85, 89, 90, 122, 124, 159, 167, 179, 198*

た

タービン発電機軸　*165*
ダイオード　*62, 175, 179, 184, 185, 186, 187*
ダッシュ（―）　*63, 64, 65*
対　格　*207*
だ円形　*192*
対　応　*9, 13, 17, 21, 25, 26, 43, 56, 57, 90, 92, 94, 109, 116, 125, 144, 162, 172, 196, 207, 213, 217, 220, 228, 231, 232, 235*
大気放電　*170*
大規模集積回路素子　*182*
台形則　*59*
対　称　*14, 140*
対　数　*97*
体積積分　*133*
対地雷　*60, 170*
大電流　*47, 170, 179*
題　目　*2, 3, 4, 5, 6, 7, 10, 12, 16, 18, 22, 24, 74, 75, 76, 78, 116, 218*
卓上電子計算機　*182*
足し算　*98*
多重化　*182, 183, 195*
多数キャリヤ　*49, 179*
縦　軸　*83, 85, 93, 122*
単位法線　*131*

255

索　　引

単位法線ベクトル　131
短縮形　2, 37, 41, 44
炭素ポリエチレン化合物　11
単　調　17, 31, 83, 85
断　面　14, 53, 80, 121
短　絡　150
段　落　6, 10, 12, 140
短論文　213

ち, つ

チャットボット　205, 210
中　央　21, 94, 153
調速機制御　165
超々高電圧　168, 169
超伝導磁気エネルギー貯蔵　170, 173
超伝導磁石　172, 173
長方形　14, 191, 192
直線偏波　192
直　流　46, 53, 143, 151, 152, 156, 157, 166, 235
直流高電圧　166
直　列　46, 55, 63, 147, 151, 152, 154, 156, 157, 182
直列演算　182
著作権　71, 214, 220, 222
著作権で保護された図　220
著　者　3, 5, 6, 14, 23, 24, 25, 26, 27, 29, 41, 67, 71, 74, 75, 76, 113, 115, 116, 212, 218, 225, 226
著者略歴　3, 27, 71
ツェナー効果　179

て

ディジタルオシロスコープ　159, 160
ディジタル加入者回線　198, 199
ディジタル計測器　159, 160
デジタルトランスフォーメーション　208, 209
ディジタルマルチメータ　159, 160
データ通信　7, 11, 70, 80, 194
デビットカード　236
デルタ接続　148
定　格　146
定冠詞　41, 42, 43, 44, 55, 105
定　義　26, 66, 128, 130, 132, 135, 136, 146
抵　抗　20, 28, 45, 46, 47, 55, 59, 60, 63, 143, 144, 146, 147, 148, 149, 151, 152, 154, 156, 157, 159, 176, 177
抵抗器　55, 63, 144, 146, 151, 152, 157
抵抗率　176, 177
定常状態　152
停　電　165, 173
敵対的生成ネットワーク　209, 210
適　用　34, 47, 54, 55, 107, 172, 173, 185, 210
鉄　芯　63, 149
電圧形変換器　166
電圧計　155, 156, 157
電圧低下　160, 167
電　位　49, 132, 134, 136, 150, 156, 177, 179
電　荷　21, 48, 49, 65, 66, 90, 132, 133, 136, 137, 138, 139, 140, 170, 177, 179
展　開　141, 142, 198
電　界　21, 31, 48, 50, 65, 66, 90, 132, 133, 138, 140, 142, 143
電荷担体　177, 179
電荷保存　137, 139, 140
電　気　6, 27, 29, 58, 62, 66, 67, 75, 78, 113, 115, 116, 121, 123, 128, 143, 145, 146, 149, 150, 162, 163, 165, 168, 173, 180, 216, 217, 220, 225, 226, 227, 228, 229, 235, 236, 238, 239
電気回路　143, 149

電気学会　217, 238
電源コンセント　235
電　子　6, 29, 58, 68, 75, 113, 115, 117, 124, 153, 158, 159, 175, 177, 179, 182, 184, 194, 198, 199, 201, 213, 214, 215, 216, 219, 220, 223, 224, 225, 226, 227, 228, 235, 237, 238
電磁エネルギー　153
電磁界　47, 52, 55, 68, 113, 138, 139, 142, 189, 191, 192, 228
電子デバイス　175, 184
電磁波　21, 137, 139, 190, 192
電子ビーム　159
電子メール　68, 198, 213, 214, 215, 219, 220, 223, 225, 226, 227, 237
伝送線路　46, 49, 51, 150, 153, 154
電　卓　50, 182, 187
電　池　157, 170, 172, 173, 174, 175, 179, 235
電池駆動時間　174
点電荷　49
伝　導　20, 21, 23, 86, 91, 122, 170, 172, 173, 177
伝導電流　21, 86
伝搬特性　8, 9, 11, 12, 22, 29, 33, 77, 78, 80, 91, 121
電流計　47, 155, 156, 157
電流密度　135, 136, 139
電力系統　164, 165, 166, 174, 175
電力ケーブル　3, 4, 7, 9, 10, 11, 12, 14, 16, 17, 18, 20, 21, 22, 23, 29, 31, 33, 53, 70, 75, 76, 77, 78, 80, 81, 83, 86, 87, 91, 93, 109, 121, 122, 123
電力システム　160, 170
電力線搬送通信　3, 4, 7, 8, 9, 10, 11, 12, 16, 18, 22, 64, 75, 76, 77, 80

電力流通網　*161, 162*
電話機　*185, 186*
電話線　*153*

と

ドーピング　*177*
トランジスタ　*53, 58, 61, 175, 180, 181, 182, 183*
ドリフト　*49, 179*
等位接続詞　*45, 46, 47, 67*
等　価　*11, 29, 34, 35, 36*
透過深さ　*8, 21, 23, 88, 89, 122*
同　期　*195, 196, 198, 199*
同期化マルチメディア統合言語　*198, 199*
投　稿　*2, 4, 6, 12, 23, 43, 62, 68, 71, 72, 212, 213, 214, 215, 216, 219*
投稿規定　*2, 62, 212, 213*
等差数列　*102*
当座預金口座　*236*
同　軸　*46, 51, 90, 153*
頭字語　*2, 4, 6, 9, 42, 55, 78, 210*
導　出　*23, 60*
透磁率　*21*
導　体　*11, 14, 17, 21, 30, 31, 45, 51, 53, 57, 59, 60, 62, 65, 66, 70, 80, 81, 83, 89, 94, 121, 122, 142, 144, 145, 153, 159, 160, 173, 175, 176, 177, 180, 188, 190*
導電率　*8, 9, 11, 14, 17, 18, 20, 21, 23, 31, 40, 57, 64, 78, 81, 83, 85, 86, 87, 88, 89, 90, 91, 93, 107, 122, 123*
導波路　*51, 153*
等比数列　*102*
透　明　*75, 159*
動名詞　*55*
特　性　*8, 9, 11, 12, 22, 29, 33, 49, 58, 62, 66, 77, 78, 80, 91, 94, 121, 123, 154, 155, 190, 192, 194*
特性インピーダンス　*49, 94, 154, 155*
匿　名　*24*
独　立　*9, 51, 57, 140, 153*
独立分詞構文　*51, 153*
特　許　*191, 192*

な

ナトリウム硫黄電池　*174*
なだれ効果　*179*
内　積　*128*
内　蔵　*43, 182, 183*
内部インピーダンス　*150*

に

ニューラルネットワーク　*204, 208*
二酸化炭素の排出　*162, 163*
二次元円筒座標系　*12, 14*
2重積分　*101*
二重母音　*207*
2進化10進数　*182, 183*
似　た　*56, 188, 236*
2年ごとに開催される会議　*214*
日本学術振興会　*27, 230*
入　射　*154, 155*

ね

ネットワーク管理者　*200, 201*
ネットワークトポロジー　*195*
熱膨張　*11, 70, 80*

の

ノートパソコン　*43, 185, 188, 235*
のこぎり波発生器　*158, 159*
能動態　*7, 27, 28, 29*

は

ハーバード方式　*25, 26, 67*
ハイフン（-）　*63, 64, 95*
バイポーラ接合トランジスタ
　　53, 175, 180, 181
パソコン　*15, 37, 43, 108, 112, 125, 185, 188, 215, 235*
バックナンバー　*223, 224*
パルス　*17, 46, 55, 83, 85, 93, 107*
バンクーバー方式　*25, 26*
背　景　*9, 10, 74, 77, 79*
配電線　*145, 153*
博士研究員　*227, 229*
博士号　*27, 39, 228, 232*
博士論文　*232*
白熱電球　*185, 186, 188*
波　形　*17, 30, 43, 57, 83, 86, 87, 93, 122, 158*
派　生　*207*
破　線　*93*
波　長　*188, 191, 192, 195*
発光ダイオード　*62, 184, 185, 186, 187*
発　散　*135, 136, 139, 140, 142*
発電所　*53, 162, 163, 164, 165, 166, 172, 173, 174, 188*
範　囲　*8, 11, 12, 14, 21, 65, 72, 81, 85, 89, 94, 122, 139, 157, 164, 165, 166, 172, 173, 177, 191, 192, 194, 232*
半教師あり学習　*203, 204, 208*
半　径　*8, 9, 14, 20, 21, 23, 31, 81, 86, 88, 91, 94, 122*
半径方向　*8, 9, 20, 21, 23, 86, 88, 91, 122*
反磁性　*149*
反　射　*154, 155*
半導体　*53, 57, 62, 66, 159, 160, 175, 176, 177, 178, 180, 188*
半導体バンドギャップ電圧基準　*159, 160*
半導電層　*3, 4, 7, 8, 9, 10, 11, 12, 14, 16, 17, 18, 20, 21, 22, 23, 29, 31, 64, 70, 75, 76, 77, 78, 80, 81, 83, 85, 86, 87, 88, 89, 90, 91, 93,*

索　引

107, 121, 122, 123
反発　179
反比例　63, 143

ひ

ビザ　215, 226, 233, 234, 237
ヒ素　177
比較　18, 43, 55, 56, 62, 70, 88, 103, 109, 112, 159, 160, 192, 216, 228
光　14, 58, 61, 62, 139, 140, 170, 184, 185, 186, 187, 188, 189, 192, 193, 194, 195, 196
光回線終端装置　196
光電子デバイス　184
光ファイバ　193, 194, 195
引き算　98
皮相電力　146, 147
左から掛ける　130
非同期転送モード　195
微分　100, 101, 129, 130, 131, 142
比誘電率　14, 20, 21, 81, 89, 122
表　3, 15, 25, 26, 44, 74, 75, 92, 94, 99, 100, 144, 212, 214, 233, 234, 235, 236
評価　6, 17, 25, 164, 212, 216
剽窃　68, 71
費用対効果　194
表の行　94
表の列　94
表皮効果　60
表面　11, 14, 21, 48, 50, 70, 80, 89, 131, 133, 142, 173
表面積分　50, 133, 142
比例　63, 132, 143, 173

ふ

プラグインハイブリッド電気自動車　162, 163
プランク定数　44
プリント　123, 182, 183

プリント回路基板　182, 183
ブルーブック　237
ブロードバンド　194, 195
プログラム　14, 23, 59, 182, 190, 202, 205, 209, 210, 215, 233
プロシーディングス　2
プロセス制御装置　167
プロセッサ　14, 108, 181, 183, 184
不一致　58
風力発電所　164, 165, 166, 174
付加記合　207
深さ　8, 21, 23, 88, 89, 122
複素数　104
符号分割多重化　195
不純物　177
不純物添加　177
不定冠詞　41, 43, 55
不定詞　61
不等式　103
不適切なオーサーシップ　5
振れ(計器の)　157
付録　3, 23, 212, 214, 216, 218, 219, 233, 238
分散発電　162, 163
分子(分数の)　96, 97
分詞構文　45, 49, 50, 51, 133, 153
分数　96, 97
分母　97
分流器　157
分路リアクトル　168, 169

へ

ベクトル　18, 47, 66, 105, 128, 129, 130, 131, 132, 135, 136, 137, 139, 141, 142, 143, 149
ベクトル積　128, 129
ベクトル微分演算子　130
ベクトルポテンシャル　132, 135, 136
ヘルムホルツの定理　132, 135, 136
平方根　89, 156
並列　46, 47, 55, 63, 146, 147, 156, 157, 164, 194
変圧器　146, 148, 149, 174
変位電流　21, 86, 137, 139, 140
編集委員　213, 219
編集長　213, 214, 219
変電所　162, 163, 174

ほ

ポインティングベクトル　137, 142, 143
ポイントツーポイント　195
ポイントツーマルチポイント　195
ホウ素　177
ホームエネルギー管理システム　162, 163
ホール　177
ポスター発表　118
母音　42, 44, 207
棒グラフ　92
法線方向成分　131
放電　5, 8, 20, 23, 32, 33, 41, 43, 44, 52, 78, 86, 88, 91, 122, 167, 170, 173, 228
方法　3, 12, 13, 30, 31, 34, 43, 47, 52, 54, 58, 61, 71, 74, 80, 133, 162, 182, 209, 233, 236
星形接続　148
保証金　236

ま

マイクロストリップアンテナ　189, 191, 192
マクスウェル方程式　44, 60, 137, 139, 140, 228
まとめ　3, 22, 33, 74, 78, 90
間違い　34, 36, 38, 60, 69, 81
末端使用者　164, 195
丸括弧　130

索　引

み, む

ミリアンペア計　*157*
未来時制　*30, 32*
無限大　*156*
無効電力　*146, 147, 166*
無生物主語　*29, 163*
無線周波数　*153*
無線通信　*189*

め

メーカ　*172, 173*
メカニズム　*5, 11, 33*
Meta 社　*209*
メモリ　*14, 73, 108, 159, 182, 183, 215*
メモリ素子　*182*

も

モーメント法　*56, 68, 190*
目　的　*3, 9, 10, 11, 12, 28, 35, 40, 49, 51, 52, 55, 60, 61, 68, 69, 74, 77, 78, 79, 80, 121, 124, 125, 162, 166, 169, 199, 202, 205, 229*
最も左側　*94*

最も右側　*94*
模倣する　*204*
文部科学省　*24*

ゆ

誘電体　*66, 145, 191, 192*
誘電率　*14, 20, 21, 81, 89, 122, 192*
有毒水銀　*186*
有　用　*34, 60, 85, 238*
USB メモリスティック　*75, 215*

よ

要　旨　*2, 3, 6, 7, 8, 9, 213, 214*
洋　上　*166*
揚水式水力発電　*172, 173*
横　軸　*83, 85, 93, 122*
読出し専用メモリ　*182*
4 面体　*177*

ら

ラプラシアン　*136*
ラプラス方程式　*136*
雷　雲　*5, 32, 33, 48, 170*
雷放電　*5, 32, 33, 167, 170*
乱層雲　*170*

り, る

リ　ン　*177*
力　率　*143, 146, 147*
流　入　*131, 143, 151*
利用者指向のソフトウェア　*190*
ルーメン　*185, 186, 188*
累　乗　*81, 97, 98*

れ

レーザ　*76, 92, 94, 125, 194*
レター　*213*
冷却装置　*173, 174*
連結的な　*207*

ろ

論文募集要項　*214, 215*
論理素子　*182*

わ

ワールドワイドウェブコンソーシアム　*198, 199*
ワット　*146, 168, 173, 174, 185, 188*
割り算　*99*

著者および監修者紹介

馬場吉弘（Yoshihiro Baba）

1971年生まれ．1994年東京大学卒業．1999年博士（工学）．1999年同志社大学助手．2003年から2004年まで米国フロリダ大学客員研究員．2012年から同志社大学教授．電磁界パルスと電磁両立性に関する研究に従事．1999年および2014年電気学会学術振興賞論文賞，2011年および2024年同進歩賞受賞，2014年および2023年同著作賞受賞．2014年IEEE EMC部門 Technical Achievement Award受賞．2009年から2018年までIEEE Transactions on Power Deliveryエディタ．2022年からIEEE Transactions on EMCアソシエイトエディタ．IEEEフェロー，IETフェロー，電気学会フェロー．

著書：Electromagnetic Computation Methods for Lightning Surge Protection Studies（Y. Baba, and V. A. Rakov, Wiley/IEEE Press, 2016），Lightning-Induced Effects in Electrical and Telecommunication Systems（Y. Baba, and V. A. Rakov, IET, 2020），過渡現象論（馬場吉弘，数理工学社，2022）．

William A. Chisholm

1955年アメリカニューヨーク州プラッツバーグ生まれ．1977年トロント大学卒業．1983年Ph.D.（ウォータールー大学）．1976年から2007年までキネクトリクス社（旧オンタリオハイドロ社の研究部門）勤務．その後1年間ケベック大学シクーティミ校教授．電力設備の雷防護および電力系統の絶縁に関する研究に従事．2001から2005年までIEEE Transactions on Power Deliveryエディタ．2007年IEEEフェロー．

著書：Insulators for Icing and Polluted Environments（M. Farzaneh, and W. A. Chisholm, Wiley/IEEE Press, 2009），Electrical Design for Overhead Power Transmission Lines（M. Farzaneh, S. Farokhi, and W. A. Chisholm, McGraw-Hill, 2012），他．

電気電子系学生のための英語処方　改訂版
論文執筆から口頭発表のテクニックまで

2013 年 1 月 30 日	初　版	1 刷発行
2022 年 9 月 15 日		3 刷発行
2024 年 12 月 15 日	改訂版	1 刷発行

発行者	本 吉 高 行
発行所	一般社団法人　電 気 学 会 〒102-0076 東京都千代田区五番町 6-2 電話(03)3221-7275 https://www.iee.jp
発売元	株式会社　オーム社 〒101-8460 東京都千代田区神田錦町 3-1 電話(03)3233-0641
印刷所 製本所	株式会社　太平印刷社

落丁・乱丁の際はお取替いたします　　　　Ⓒ2024 Japan by Denki-gakkai
ISBN 978-4-88686-324-9　C3050　　　　　Printed in Japan

電気学会の出版事業について

　電気学会は，1888年に「電気に関する研究と進歩とその成果の普及を図り，もって学術の発展と文化の向上に寄与する」ことを目的に創立され，教育関係者，研究者，技術者および関係諸機関・法人などにより組織され運営される公益法人です．電気学会の出版事業は，1950年に大学講座シリーズとして発行した電気工学の教科書をはじめとし半世紀以上を経た今日まで電子工学を包含した数多くの図書の企画，出版を行っています．

　電気学会の扱う分野は電気工学に留まらず，エネルギー，システム，コンピュータ，通信，制御，機械，医療，材料，輸送，計測など多くの工学分野に密接に関係し，工学全般にとって必要不可欠の領域となっています．しかも年々学術，技術の進歩が加速的に速くなっているため，大学，高専などの教育現場においては，教育科目，内容，授業形態などが急激に様変わりしており，カリキュラムも多様化しています．

　電気学会では，そのような実情，社会ニーズなどを調査，分析して時代に即応した教科書の出版を行っていますが，さらに，学問や技術の進歩に一早く応えた研究者，エンジニア向けの専門工学書，また，難解な専門工学を分りやすく解説した一般の読者向けの技術啓発書などの出版にも鋭意，力を注いでいます．こうしたことは，本学会が各界の一線で活躍する教育関係者，研究者，技術者などで組織する学術団体だからこそ出来ることです．電気学会では，これらの特徴を活かして，これからも知識向上，自己啓発，生涯教育などに貢献できる図書を出版していきたいと考えています．

会員入会のご案内

　電気学会では，世代を超えて多くの方々の入会をお待ちしておりますが，特に，次の世代を担う若い学生，研究者，エンジニアの方々の入会を歓迎いたします．電気電子工学を幅広く捉え将来の活躍の場を見出すため入会され，最新の学術や技術を身につけ一層磨きをかけてキャリアアップを目指してはいかがでしょうか．すべての会員には，毎月発行する電気学会誌の配布など，いろいろな特典がございますので，是非一度下記までお問合せ下さい．

〒102-0076　東京都千代田区五番町6-2　一般社団法人　電気学会
　https://www.iee.jp　Fax：03(3221)3704
　▽入会案内：総務課　Tel：03(3221)7312
　▽出版案内：編修出版課　Tel：03(3221)7275